The Compressible Fluid Physics of the Human Respiratory System

The Compressible Fluid Physics of the Human Respiratory System provides a comprehensive exploration of the principles of fluid dynamics, classical mechanics, and their applications to complex systems such as human respiration.

It dives into foundational concepts like Newtonian mechanics and stress-strain relationships while also addressing advanced topics such as compressible fluid flow, viscosity, and energy conservation. It uniquely combines classical principles with real-world phenomena, such as the behavior of gases, liquids, and solids under various forces and pressures. The inclusion of detailed mathematical derivations, historical context, and empirical models adds depth to the discussion. Its interdisciplinary approach makes it valuable for readers interested in physics, engineering, and biological sciences.

This book is designed for a general audience with an interest in science and medicine irrespective of mathematical background.

Foundations of Biochemistry and Biophysics

This textbook series focuses on foundational principles and experimental approaches across all areas of biological physics, covering core subjects in a modern biophysics curriculum. Individual titles address such topics as molecular biophysics, statistical biophysics, molecular modeling, single-molecule biophysics, and chemical biophysics. It is aimed at advanced undergraduate- and graduate-level curricula at the intersection of biological and physical sciences. The goal of the series is to facilitate interdisciplinary research by training biologists and biochemists in quantitative aspects of modern biomedical research and to teach key biological principles to students in physical sciences and engineering.

Authors are also welcome to contact the publisher (Physics Editor, Carolina Antunes: carolina.antunes@tandf.co.uk) to discuss new title ideas.

Light Harvesting in Photosynthesis
Roberta Croce, Rienk van Grondelle, Herbert van Amerongen, Ivo van Stokkum (Eds.)

An Introduction to Single Molecule Biophysics
Yuri L. Lyubchenko (Ed.)

Biomolecular Kinetics: A Step-by-Step Guide
Clive R. Bagshaw

Biomolecular Thermodynamics: From Theory to Application
Douglas E. Barrick

Quantitative Understanding of Biosystems: An Introduction to Biophysics
Thomas M. Nordlund

Quantitative Understanding of Biosystems: An Introduction to Biophysics, Second Edition
Thomas M. Nordlund, Peter M. Hoffmann

Entropy and Free Energy in Structural Biology: Thermodynamics, Statistical Mechanics and Computer Simulation
Hagai Meirovitch

Metabolism and Medicine: Two Volume Set
Brian Fertig

Introductory Science of Alcoholic Beverages: Beer, Wine, and Spirits
Masaru Kuno

The Compressible Fluid Physics of the Human Respiratory System
W. R. Matson

https://www.crcpress.com/Foundations-of-Biochemistry-and-Biophysics/book-series/CRCFOUBIOPHY

The Compressible Fluid Physics of the Human Respiratory System

W. R. Matson

CRC Press
Taylor & Francis Group
Boca Raton London New York

CRC Press is an imprint of the
Taylor & Francis Group, an **informa** business

First edition published 2026
by CRC Press
2385 NW Executive Center Drive, Suite 320, Boca Raton FL 33431

and by CRC Press
4 Park Square, Milton Park, Abingdon, Oxon, OX14 4RN

CRC Press is an imprint of Taylor & Francis Group, LLC

ISBN: 978-1-041-03124-6 (hbk)
ISBN: 978-1-041-16234-6 (pbk)
ISBN: 978-1-003-68347-6 (ebk)

DOI: 10.1201/9781003683476

Typeset in Nimbus Roman font
by KnowledgeWorks Global Ltd.

*Publisher's note:*This book has been prepared from camera-ready copy provided by the authors.

Contents

List of Figures

Preface

Thank you for reading this rather unusual book about the human respiratory system. Unlike most biological and medical texts, this one will consider the physics involved in the process of breathing, including the pressures, forces, and other parameters necessary to make breathing possible. We will also examine to some extent why breathing is necessary and how it fits in with other vital systems. We will examine the fluid dynamics involved in the operation and maintenance of this very critical system. We will also look at some non-essential systems in order to put this one process into its proper context. We will also examine the consequences of various types of damage, agitation, and illness and their impact on the respiratory system in order to get an understanding of just how these issues alter the original parameters and operational abilities of various parts of the system and the consequences that result.

Although some of the analysis can get quite involved and use some fairly advanced mathematical concepts and techniques, this is primarily an introductory-level text directed toward a general audience. Its primary purpose is not to prepare one for medical school or passing the qualifications examination to become a respiratory therapist. There are thousands of textbooks and guidebooks out there for that purpose. This text is directed toward the common person with some basic understanding of math and an acquaintance with fundamental physics who wants to understand their respiratory system a little better. Maybe they have a friend or family member undergoing some treatment or who has some kind of lung disease they don't understand. Maybe their mother or father is currently in the hospital on a ventilator, and they would like to understand what the heck the doctor or therapist was just talking about. Whatever the case, this is not a medical textbook, it is a discussion surrounding a vital bodily system that is easily damaged and often suffers seasonal challenges, and I hope to provide you, the reader, with some deeper insights into how and why the system works.

This is neither a medical nor biological textbook, and it is not intended to replace the one selected by your professor or teaching physician; however, it may serve in some capacity as a supplement but little more. In contrast with these other types of texts, it has quite a different focus and intention. As a physicist, it is my intention to direct the book's attention more to the physics of respiration rather than the medical and anatomical aspects; however, some of those issues will need to be addressed as a matter of course in order to understand the context within which these physical processes are taking place and why. Consequently, there will be times where a great deal of attention must be paid to physical dimensions and parameters as they apply to the underlying physical principles and act as constraints, issues that are not commonly addressed in other texts. There will also be several instances where more of an engineering aspect will be required, since fluid dynamics is a hybrid field of physics and engineering. Nevertheless, the attention will be mostly on the physics and mechanics of breathing, not on the medicine or biology.

I was driven to writing this book after spending five years helping my wife wrestle with various types of post-operative respiratory issues that resulted during recovery in the ICU and kept coming back over and over after discharge. I had spent nearly eighteen years in emergency medicine as a first responder before returning to school and getting a Ph.D. in experimental applied fluid physics, so I already knew something about the respiratory system; however, I found myself searching for more and more information. As a scientist, the explanations were often not sufficient. I wanted to know more, and more importantly I wanted to know why, and doctors were often too busy to answer or ill equipped to explain the situations to my satisfaction. Medicine is not necessarily about the "why." Knowing that this enzyme reacts with that protein and produces this other thing is good enough for them but not for me.

As a result of these challenges, I spent years researching into the various functions and malfunctions of the respiratory system in order to augment and expand my knowledge on the anatomy, physiology, and physics of respiration. Over these years I also found a couple of specialists in pulmonology and otolaryngology that were actually willing to talk to a Ph.D. physicist for a while, and I believe we both found those conversations quite illuminating.

When I was approached about writing a third book, this five-year journey was coming to a close. Thanks to the teamwork I shared with over a dozen physicians in multiple states, the serious respiratory challenges are finally in the past, and I have compiled this work in order to pass on this information to others who might share a similar desire for answers and information. These are the explanations and interpretations of an empirical physicist, not a medical doctor, so they will no doubt be a bit different when it comes to the viewpoint and perception you might get from a physician, nurse, or therapist.

I hope you enjoy this book. Please feel free to explore my publications on other topics where physics and life intersect. My intended audience is always you, the people, so I am sure you will find them both accessible and enjoyable. In the meantime, enjoy this book. It is my hope that you will find it both reachable and interesting in every possible meaning of that word.

About the Author

Dr. W. R. Matson joined the fire service in 1982 and received his initial emergency medical certification in 1984. He spent the next ten years seeking additional medical training and certifications and serving as a certification instructor for American Heart Association (AHA) and the Red Cross until an industrial incident ended his service in 1997, after which he returned to school and finished his bachelor's degree in physics in 1999 and doctoral degree in fluid physics in 2004. He has since served as a professor of physics at several universities.

1

Introduction to Key Life Systems

The human body is the most sophisticated and complex machine in the known universe. Every second, thousands of processes and reactions take place, millions of signals are sent and received, and terabytes of data are analyzed by the most complex processing system in existence. All of these processes and reactions are carefully and artfully managed and controlled through a highly intricate electrical network unlike anything ever created by humans and remains far beyond our ability to reproduce, and most of this is done automatically without the need for conscious control or the slightest awareness on our part.

The human brain has a hundred billion neurons, each connected to several thousand other neurons, in an intricate and highly complex network that allows the processing, manipulation, and storage of information. The human brain is arguably the most complex object in the known universe.

> If the human brain were so simple that we could understand it, we would be so simple that we could not. Thankfully, the complexity of our brain is so great that we are not simple and neither, therefor, is the task of understanding it.
>
> Emerson Pugh (1929-2024)
> Former president IEEE

DOI: 10.1201/9781003683476-1

1.1   A Little Background

Our understanding of bodily systems stretches back to ancient times. Early civilizations in Egypt, Mesopotamia, and Greece observed and documented the anatomy of humans and other animals mainly through dissection, for example, Herophilus who worked in Alexandria. Egyptians were keenly interested in injuries and how to treat them. Greek physicians like Aristotle, Alcmaeon of Croton, and Hippocrates studied not only the body's structure but also how it functions. Claudius Galen (129–216) is commonly recognized for developing the experimental methods used in medical investigations.

In the 16th century, Andreas Vesalius (1514–1564) shown in Figure 1.1 published detailed accounts of various anatomical

FIGURE 1.1

A portrait of Andreas Vesalius (1514–1564) attributed to Jan van Calcar in 1543. Vesalius published detailed accounts of various anatomical studies. His most notable publication was *De Humani Corporis Fabrica,* and this portrait was featured in that text.

FIGURE 1.2

German physician Rudolf Virchow (1821–1902) took a special interest in anthropology and pathology. The invention of the microscope
opened the doors to the study of cells, tissues, and pathogens, and
Virchow became one of the leading researchers in examining diseases
at the cellular level. (National Institutes of Health, part of the United
States Department of Health & Human Services.)

studies. His most notable publication was *De Humani Corporis Fabrica.* Leonardo da Vinci also made and recorded detailed observations
with meticulous precision. The 18th century saw the invention of the
microscope and opened the doors to the study of cells, tissues, and
pathogens which, for the first time, allowed for the explanation of diseases at the cellular level. One of the leading researchers to take advan-
tage of this development was Rudolf Virchow (1821–1902), shown in
Figure 1.2, a German physician with a special interest in anthropology
and pathology.

The 19th century also saw huge strides in the field of medicine,
understanding, and practice. Anesthesia was introduced along with

several improvements in surgical techniques, including Joseph Lister's introduction of antiseptic practices using carbolic acid to sterilize both instruments and wounds. This resulted in significantly fewer post-operative infections and complications. Louis Pasteur and Robert Koch introduced the germ theory of diseases and demonstrated that many illnesses were caused by microorganisms. This laid the groundwork for future developments in vaccines and antibiotics. Edward Jenner developed a vaccine for smallpox that was a landmark achievement. Later, Pasteur would develop vaccines for cholera and rabies. Rene Laennec invented the stethoscope which was able to allow doctors to listen to a patient's heart and lungs. Semmelweis actually discovered that washing hands with chlorinated lime would prevent disease transmission and highlighted the importance of hygiene in the practice of medicine. Thomas Rocyn Jones, along with other orthopedic physicians, made significant advancements in the treatment of fractures, dislocations, and muscle injuries, including the invention of the first splints. James Blundell successfully administered the first blood transfusion, a practice that is almost commonplace today.

The 20th century had a hard act to follow, but it failed to disappoint by marking some of the most significant and transformative breakthroughs in human history, starting with the discovery of penicillin in 1928 by Alexander Flemming, shown in Figure 1.3. Jonas Salk developed a vaccine for polio in 1955 following the first successful kidney transplant, or transplant of any kind, in 1954 by Joseph Murray. The success is largely accredited to advancements in immunosuppression in order to prevent rejection and organ perfusion systems to maintain organ viability during transfer. Since then, transplants have included liver, heart, lungs, pancreas, eyes, and other organs. The first X-rays under clinical conditions were used by John Hall-Edwards in Birmingham, England in 1896. Since then, computed tomography (CT) scans were put into practice in 1971 thanks to the significant advancements in computer processing abilities of that era. These enhanced images are built up of numerous X-rays, which are captured as slices and can be combined to produce three-dimensional images of organs, bones, and other structures. X-rays saw further, and possibly final, enhancement in 2000 with the development of the digital detection plate instead of using film. Developed in the 1990s, this technology allowed for post-processing

FIGURE 1.3
Professor Alexander Fleming at work in his laboratory at St Mary's Hospital, London, during the Second World War. Professor Fleming was a Scottish biologist and physician. In 1928, he noticed a particular mold growing in a petri dish inhibited the growth of other bacteria. He identified the mold as Penicillium Notatum and was able to isolate and adapt the active compound which he then named Penicillin. (Ministry of Information Photo Division Photographer. Scanned and released by the Imperial War Museum, United Kingdom Government.)

and a much lower signal intensity, reducing the radiation dose, thereby allowing patients to undergo more frequent examinations. This laid the groundwork for the mammogram campaign to promote early detection of breast cancer. The MRI became commercially available in 1980 and presented the first neutral-impact diagnostic three-dimensional imaging device with comparable or superior detection and resolution as can be seen in the side-by-side comparison in Figure 1.4. Unlike X-rays, the magnetic fields are harmless to the human body, but just like the CT, the images are made up of numerous slices to give the three-dimensional

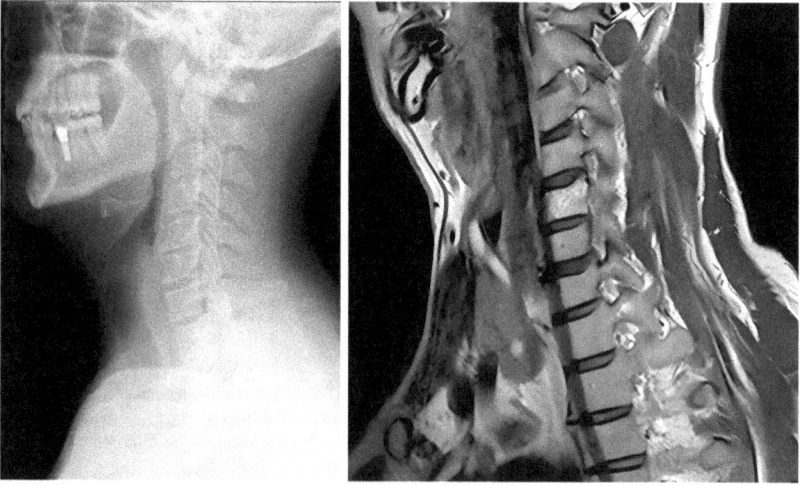

FIGURE 1.4
Side-by-side comparison of a cervical X-ray and a cervical MRI. While the X-ray is faster and very good at showing bone structure, notice the enhanced detail in the MRI. However, this comes with a reduced field of view, and it is often necessary to compare numerous "slices" to get a good three-dimensional impression.

perspective. Ultrasound was first discovered in the 1800s, but it was not until the 1920s that its use as a medical imaging system was proposed. It was another 40 years until ultrasound devices became readily available in the 1960s. Since then, ultrasound devices have undergone continued improvement, including real-time display and processing, resolution, and measurement capabilities. The concept of positron emission tomography (PET) was first proposed in 1951, but it was not until the 1970s that its use as a diagnostic instrument was made clear. They are most useful in the location, identification, and tracking of tumors and certain neurological conditions, such as Alzheimer's, Parkinson's, and epilepsy.

In contrast with the three previous centuries, the 21st century is off to a slow start. Thus far, the major advancements of this century include telemedicine, electronic health record-keeping, the ability to wear diagnostic devices, and virtual reality headsets with earphones. There are a great many diseases and conditions that have yet to be understood, treatment methods or abilities that need to be developed,

FIGURE 1.5
Alois Alzheimer was a German psychiatrist and neuropathologist. In 1901, he observed a 51-year-old woman exhibiting symptoms of memory loss, confusion, and paranoia. After her death, he examined her brain and found abnormal deposits of proteins and large collections of tangled nerve cells. These conditions are now associated with Alzheimer's Disease which effects more than half a million additional people with each passing year. (National Library of Medicine, National Institutes of Health, part of the United States Department of Health & Human Services.)

and challenges to be overcome. For example, *Fibrodysplasia Ossificans Progressiva* is a rare genetic disorder where muscles and tendons turn into bone. Older and more well-known conditions remain a challenge, such as Alzheimer's. According to a recent article,[1] Alzheimer's drugs fail 99.6% of the time which is 47 times worse than cancer treatments, even though Alzheimer's disease was first identified more than a century ago in 1906 by the German psychiatrist and neuropathologist Alois Alzheimer (1864–1915) shown in Figure 1.5. After more than a century,

---

[1] Twenty-first century diseases: Commonly rare and rarely common?

the exact causes of the tangled nerve cells and protein deposits in the brain still have not been explained; furthermore, there is no single, definitive, and objective diagnostic test that can even diagnose the disease prior to an autopsy, though PET scans show some promise.

While we still have a long way to go to fully understand the machines that we are, progress made over the previous three centuries has extended our expected lifetimes and greatly improved the quality of whatever lifespan we might realize. We can only hope that research into the mysteries of old ultimately bears fruit and that technological developments open new doors and opportunities in the future.

1.2   Major versus Minor Systems

The abilities and systems of the human body can be broken down into two parts: those that are required for continued survival and those that contribute mainly to the quality of life and are, to some extent, optional from the standpoint of remaining alive. Essential functions include breathing, circulation, digestion, and regulation. Optional functions include walking, talking, climbing, vision, hearing, memories, and so on.

The "simple" act of walking has troubled robotics engineers for decades. This is why most operational robots have four legs, so that at least two of them are always on the ground at any given moment of time. The process of bipedal walking requires timing and balance, which is why stroke victims have such noticeable walking styles, called *gaits,* that very nearly mimic that of our four-legged robots. To put it bluntly, walking is a rhythmic process of constantly falling and catching yourself over and over. Imagine for a moment what would happen if you were walking along and forgot to put your other foot down in time. Consider what happens when your foot comes down on a very slippery surface. This is evidence of the repetitive falling and catching process. This complex pattern of behavior is something we do habitually without thinking, because we have done it since we were but a few years of age, but it is nearly impossible for most robots to achieve. A robot has no sense of balance, and simply preprogramming a rhythmic gait is not sufficient. We have to change our pattern constantly to account for obstacles, changes of direction, changes of inclination, changes of surface conditions, and so on. We are constantly compensating for the world around us.

How we perceive our environment is equally as amazing. The human eye is capable of recording a full 200° field of view in a staggering 120-megapixel full-color image capable of perceiving more than one million colors and hundreds of shades of gray. All of this is provided at the rate of between 30 and 120 frames per second, [2] depending on what we are looking at and our immediate attention focus. Furthermore, we do not have only *one* of these amazing devices; most of us are actually equipped with *two* of them which allows for triangulation and depth perception, as shown in Figure 1.6. Our brain, the incredible *central processing center* that it is, inherently "knows" how far apart our two eyes are at any given moment (how, we don't quite know). It also knows the current direction each eye is facing at any given instant. Using these two factors, determining the distance to a given object is elementary geometry:

$$Tan\theta = \frac{D}{L/2} = \frac{2D}{L} \tag{1.1}$$

where D is the distance to the object, L is the distance between each eye, and θ is the angle at which the eye is facing, measured such that $\theta = 90°$ is straight ahead. The calculation is then simply

$$L = \frac{2D}{Tan\theta} \tag{1.2}$$

It is interesting to note that literature on the accuracy of the angle measurement is unclear, and the actual distance estimate arrived at by the brain may include other factors such as environmental indicators. Furthermore, one may note that this calculation requires only a single eye; therefore, having the second opinion clearly helps cement an accurate distance estimate. Modern optical equipment can gather information from light sources our human eyes cannot see, and our technological resolution is slowly approaching that of the eye (currently at about

---

[2]Movie playbacks are often set to 60 *fps* because of the latency of the human eye; however, the brain can process images at speeds up to 300 *fps*. Playback rates below 30 *fps* will often generate an annoying flicker that could cause headaches or trigger epileptic seizures.

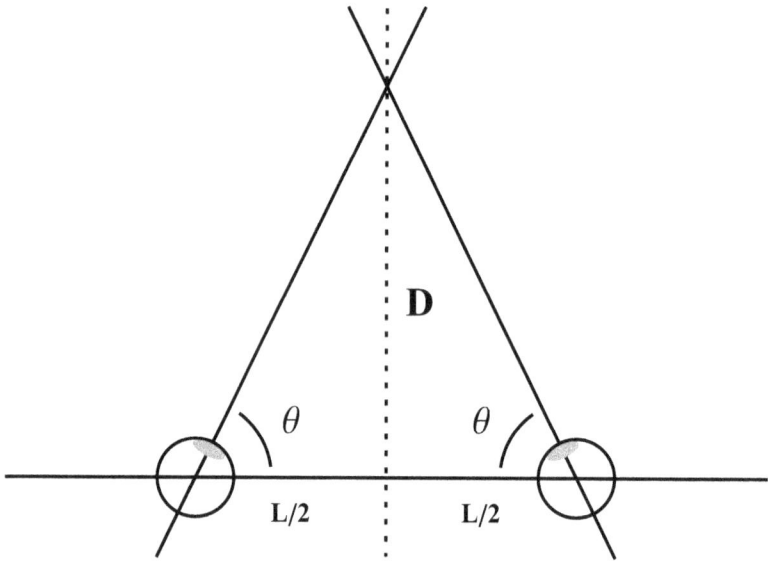

FIGURE 1.6
Using the angle of each eye and the distance between them, the distance to any given target can be obtained using the tangent of the angle of rotation and the distance to each eye as measured from body center. In this case, the angle is measured such that $90°$ is straight ahead.

10% for most equipment), but the other capabilities, including field of view and three-dimensional triangulation, are still far beyond our technological abilities at this time.

The human ear is just as amazing as the human eye. It is capable of discerning sound intensities of as little as $10^{-12} \, W/m^2$. In contrast, the wings of a butterfly generate intensities of $10^{-10} \, W/m^2$, one hundred times stronger than our ears are capable of perceiving. Furthermore, they can tolerate intensities up to $1 \, W/m^2$; however, at this point mechanical damage will start destroying the original equipment. Total destruction will occur for sounds with intensities greater than $100 \, W/m^2$. Alexander Graham Bell (1847–1922) devised a logarithmic scale which represents our perceived "loudness" of a given sound. He took the minimum intensity of $I_0 = 10^{-12} \, W/m^2$ as the threshold and scaled his measurements from that point.

$$B = Log_{10}\left(\frac{I}{I_0}\right) \qquad (1.3)$$

or

$$dB = 10\,Log_{10}\left(\frac{I}{I_0}\right) \qquad (1.4)$$

The butterflies with wing sounds of intensity $10^{-10}\,W/m^2$ would rank 20 on the *decibel* scale, written 20 *dB* in honor of Bell. As if this sensitivity were not enough, the ears can also process and discern more than 30,000 frequencies across the sensitivity range of 20 *Hz* to 20,000 *Hz*, giving a resolution of nearly 1/2 *Hz*, not accounting for interpolation, and they report these frequencies in real time to the brain. Since we have auditory receptors on each side, this also provides sonar direction finding abilities so that we know from which direction a given sound originated. Furthermore, our central processor, the brain, is also capable of "filtering" selected frequency spectra, allowing us to "selectively listen" to one voice or sound source immersed within a crowd of other voices and sounds. We have artificial sound sensors that can pick up ultra-low and ultra-high frequencies and everywhere in-between, but no single device that does *all* of this range. Furthermore, our technological abilities in processing that sound are far inferior to the real-time spectral analysis, tracking, and filtering abilities every one of us has done and continues to do naturally since birth. We are making progress with artificial cochlea and voice-recognition software, but we have a very long way to go.

1.3 Organs or Essentials

The systems discussed in the previous section might seem a bit different than systems discussed in medical texts. In this discussion, I defined a "system" as a complete set of collection, conversion, and analysis processes. The eyes, for example, would collect the light, convert it to neurological signals, and pass the information on to the optical cortex for image analysis and processing. However, from a medical viewpoint, most textbooks would discuss a system as a collection of one or more organs that would fulfill some specific task or function. While these two definitions are slightly different, in most cases these two interpretations are in complete agreement. The operational definition is simply a bit more broad and includes things like eyes and ears where, perhaps, other books would not.

While the actual number of "systems" in the human body seems to vary slightly from one text to the next, most would agree there are from 10 to 12 organ-based systems. While the focus of this text will be the respiratory system, it is important to at least introduce some of the companion elements that contribute to the overall mission of sustaining our lives such that we might put the functional operation of the respiration system into proper context and better understand its contribution to the overall mission. This will become even more relevant when we start discussing the interface between these systems under normal and abnormal conditions.

1.3.1   Integumentary

The integumentary system consists of all those external elements that surround and protect the body from the environment and its negative influences, such as bacteria and injury. The number one participant in this system is the skin, the largest organ of the body by far. The skin itself has several sub-components, including hair follicles, sweat glands (which help in temperature regulation), sebaceous glands (which produce the skin oils), and nerves to detect pressure (including touch), temperature, and pain. In addition, the system includes the nails at the end of your fingers and toes, which provide a kind of armor plating to these sensitive areas.

The primary responsibility of the skin, all of its parts, and the nails is protection. To this end, the skin is amazingly resistant and resilient compared to many other materials. It is completely waterproof and thus impervious to bacteria, viruses, and other pathogens. The outer layer, called the *epidermis,* is composed of dead cells; consequently, it also provides a kind of sacrificial layer to protect against mild acids and other abusive chemicals, scrapes, and other forms of mechanical damage. The fat layer under the skin also helps the skin provide a cushion against blunt impact, such as falls and bumps. The ability of the skin to produce melanin, a *UV*-absorbing pigment that produces the brown tan color, also helps protect the interior organs and muscles from dangerous solar radiation.

The skin also plays a participatory role with other systems by providing sensory and early warning information as well as regulatory assistance. Nerves can advise the brain about external temperatures, pressures, and contact information. By directing more or less blood flow to the outer layers, the body can either shed or retain internal heat, thereby regulating body temperature. Sweat glands kick in for a little extra boost in cooling the blood when needed. The tissues can help maintain a proper water balance so the body does not drown itself or become dehydrated. Certain skin cells, such as the Langerhans cells, can issue an emergency alert to the nervous system when they detect certain foreign entities attempting to enter the body.

The nails are attachments to the skin at precisely twenty locations: the ends of each finger, thumb, and toe. The nails grow from the nail root located a short distance beneath the skin fold that covers the top part of the nail called the *eponychium*. The small white part that you see right next to the skin is the *lunula* from the word *luna* meaning moon, since it looks like a small crescent moon. The bulk of the nail is the nail body.

Nails serve primarily for protection, providing a hard armor plating that covers the end of our digits. This protects our fingers and toes while we work and play. To a lesser extent, they also provide some mechanical assistance by providing a hard and relatively small tool for picking up things and scratching that itch, and they also provide a minor form of self-defense.

1.3.2 Skeletal

The skeletal system provides both shape and structure to the body. The adult human skeleton is made up of 206 bones, including 29 bones in the skull or head including the ears, 25 bones in the chest including ribs and the sternum, 26 bones in the vertebral column stretching from the head to the tailbone or *coccyx,* 6 bones in both arms and 54 in both hands, 2 bones in the pelvis or hips, 8 bones in each leg and 52 in each foot, and finally 4 in the shoulders.

The primary function of the skeletal system is fairly obvious, providing the structural framework that makes humans look the way they do; however, the skeleton provides additional services as well. The 22 bones that make up the skull surround the brain and provide protection to the most important organ in the body. The 25 bones of the rib cage protect the lungs and create an air-tight environment, which facilitates breathing. Furthermore, the marrow inside all 206 bones is primarily responsible for the production of red blood cells. The bones also serve as storage facilities for certain minerals, such as calcium and phosphate.

The bones also make it possible for us to move. The bone shafts provide the lever arms, and the joints provide the fulcrums as we lift and carry things, stand up and walk, or even crawl along the floor. Without these rigid anchors and levers, and without the smooth glide surfaces between them, walking and talking and running and crawling would be impossible.

The skeletal system typically includes the ligaments that hold the bones together and may also include the tendons, which attach muscle to bone. Ligaments are the small fibrous materials that connect one bone to another, particularly in the regions of joints. The knee, for example, has four main ligaments: the anterior cruciate (ACL), posterior cruciate (PCL), medial collateral (MCL), and lateral collateral (LCL). The shoulder actually has 6 ligaments, including the glenohumeral, coracohumeral, coracoclavicular, acromioclavicular, coracoacromial, and transverse humeral. Ligaments play a crucial function in providing both stability and function for any joint of which they are a part. They regulate the spacing and tension in the joint while also providing the flexibility for the joint to function.

1.3.3 Muscular

When one thinks of the muscular system, one usually thinks of the ability to move. While this is true, muscles provide the forces necessary to stand, walk, lift your arms, play the piano, and so on, muscles also provide shape and perform various other services.

The body literally has hundreds of muscles. Some of these are members of groups and work in tandem to provide cross-bracing and controlled movement. For example, the back has 40 muscles, each grouped into left and right pairs in order to provide stability and control. The outermost muscles provide movement of the upper back and shoulders. The intermediate muscles primarily help with breathing. And the deep muscles stabilize and support the spine.

Not all muscles are the same kind of material. The muscles found in the back, arms, and legs are skeletal muscles and provide mostly linear forces. For this reason, these muscles appear and feel very striated along the line on which they can act. Most of these muscles are also under your direct and conscious control.

However, there are also smooth muscles such as those found in veins and arteries, as well as in and around other internal organs. These tissues are very smooth, as the name suggests, and are under control of the autonomic nervous system. In other words, they are controlled by your nervous system, but you cannot consciously control them. The walls of the veins and arteries constrict and relax automatically as part of the flow and pressure control system.

The last type of muscle material is unique to the heart and is thus called *cardiac* muscle. This material is a cross between the other two. It is striated similar to the skeletal muscles since the heart has only one job: to pump blood. However, it is also completely autonomic and is beyond your ability to directly control.

In addition to providing a force, muscles can also help with heat production and circulation. Obviously, the heart is a muscle and pumps the blood throughout the body. The arteries and veins, through the act of constriction and relaxation, aim and direct the flow where needed and adjust the total volume of the circulatory system at all times. In addition, the actions of these muscles consume sugar and produce heat with every contraction. When you are cold, your body initiates random and rapid contractions to increase the consumption rate and produce more heat; thus, you "shiver." Although heat is actually a byproduct, it is essential that your body maintain a normal operating temperature at all times. While the skin can assist in shedding excess heat, the muscles provide the ability to produce heat when you don't have enough.

1.3.4 Endocrine

As part of the regulatory, control, and communications systems of the body, there are a large number of glands and organs that pro-duce hormones in order to regulate bodily functions. While the nervous system provides rapid, electrical signals for quick responses, the endocrine system acts as a slower, yet equally vital, chemical messenger. It's a network of glands that produce and release chemical substances called hormones directly into the bloodstream. These hormones travel throughout the body acting as messengers to regulate a vast array of physiological processes, from growth and metabolism to mood and reproduction. Each hormone addresses specific key issues regarding the operation or growth of the body at large.

The pituitary gland is a small, pea-sized structure located at the base of the brain, just below the hypothalamus, and is often referred to as the "master gland." The pituitary gland is divided into two main lobes, each with distinct functions. The anterior pituitary produces and secretes a wide array of hormones that regulate other endocrine glands, including growth hormones, thyroid stimulation hormones, and adrenocorticotropic hormones to stimulate the production of cortisol, to name a few. The posterior pituitary stores and releases hormones produced by the hypothalamus, including antidiuretic hormones, to regulate water balance.

The hypothalamus is actually part of the brain and provides a crucial link between the nervous and endocrine systems. It acts as the primary control center for many endocrine functions. The hypothalamus produces hormones that regulate the anterior pituitary gland.

The thyroid is located in the neck just below the larynx and produces hormones controlling metabolism and growth, most notably thyroxine and triiodothyronine. These hormones are critical for regulating the body's metabolic rate, influencing energy production, body temperature, heart rate, and brain development. The thyroid also produces calcitonin, which helps regulate calcium levels in the blood. On the posterior surface of the thyroid are four small glands called the parathyroid which act as the primary regulator of blood calcium levels.

The adrenal glands are located atop each kidney and are once again composed of two distinct regions, each producing different hormones. The adrenal cortex produces steroid hormones such as cortisol which regulates metabolism and manages stress response, and aldosterone which regulates the electrolyte balance and blood pressure. The adrenal medulla produces catecholamines, including epinephrine (adrenaline) and norepinephrine (noradrenaline) which are responsible for the "fight-or-flight" response to stressful or dangerous situations by increasing heart rate and blood pressure and releasing glucose to feed the muscles.

The pancreas produces hormones responsible for blood glucose regulation. Insulin is secreted by beta cells and lowers blood glucose levels by promoting glucose uptake by cells and its storage as glycogen. On the other hand, glucagon is secreted by alpha cells and raises blood glucose levels by stimulating the liver to release stored glucose.

The thymus gland is located in the chest and plays a vital role in the development and maturation of T-lymphocytes, white blood cells important to the immune system. While its role diminishes with age, it is essential for early immune system development.

1.3.5 Lymphatic

The lymphatic system is a complex network of vessels, tissues, and organs that plays a crucial role in maintaining fluid balance, absorbing fats, and defending the body against disease. In addition to a drainage system, it plays an integral part of both the circulatory and immune systems.

Lymph is a clear, yellowish fluid that circulates throughout the lymphatic system and is formed from interstitial fluid that has entered the lymphatic capillaries. Lymph is similar in composition to blood plasma but contains fewer proteins and no red blood cells; however, it does carry white blood cells, fats, proteins, and waste products away from the tissues.

Lymphatic vessels make up an extensive network of thin-walled vessels that are similar to blood vessels but carry lymph instead of blood. They begin as tiny, blind-ended capillaries in the interstitial spaces of most tissues. These capillaries are highly permeable which allows interstitial fluid to easily enter. As they merge, they form larger collecting vessels, which then converge into lymphatic trunks and finally into two large collecting ducts: the right lymphatic duct which drains the right upper limb, right side of the head, and right side of the thorax into the right subclavian vein, and the thoracic duct which collects lymph from the rest of the body and empties into the left subclavian vein. Unlike the circulatory system, which has the heart as a pump, the lymphatic system relies on skeletal muscle contractions, respiratory movements, and smooth muscle contractions within the vessel walls to propel lymph forward. The network has numerous valves located within the lymphatic vessels which prevent backflow.

There are several specialized structures that contain white blood cells called lymphocytes and other immune cells that are critical for immune system function. Red bone marrow is where all the blood cells, including lymphocytes, are produced. The thymus is where T cells mature and are "educated" to recognize foreign invaders and simultaneously ignore normal cells.

Lymph nodes are small, bean-shaped organs clustered along lymphatic vessels, particularly in the neck, armpits, and groin. They act as filters, trapping foreign particles, cellular debris, and pathogens such as bacteria and viruses from the lymph. Lymphocytes within the nodes destroy these harmful substances. Lymph nodes also swell when the body is fighting off an infection as a response to the increased immune system activity.

The tonsils are a bundle of lymphoid tissue located in the pharynx. They form a protective ring that traps pathogens entering the body through the mouth and nose.

The spleen is the largest lymphoid organ and is typically located in the upper left abdomen. It filters the blood and removes obsolete or damaged red blood cells, platelets, and pathogens. It also serves as a reservoir for monocytes and thus plays a significant role in the body's immune system responses.

1.3.6   Urinary

Also known as the renal system, the urinary system is a vital biological system responsible for filtering waste products from the blood and expelling them from the body as urine. It also plays an important role in maintaining proper fluid balance, electrolyte levels, blood pressure regulation, and aiding in red blood cell production.

The kidneys are a pair of bean-shaped organs located on either side of the spine just below the rib cage. They are the primary organs of the urinary system, responsible for filtering blood, removing waste products, and producing urine. Each kidney is encased in a tough, fibrous capsule. The cortex is the outer region which contains the filtering units, and the medulla is the inner region that collects urine.

The renal pelvis is a funnel-shaped structure in the center of the kidney that collects urine from the medulla and funnels it into the ureter.

Nephrons are microscopic filters and serve as the functional units of the kidney. Each kidney has more than a million nephrons. The renal corpuscle is a capillary network where the blood is actually filtered, and the Bowman's Capsule is a cup-shaped structure that surrounds the network and collects the filtered fluid. The renal tubule is a long, convoluted tube that processes the filtered fluid arriving from the Bowman's Capsule allowing essential substances to be reabsorbed back into the blood while secreting additional waste products. It includes the proximal convoluted tubule, the loop of Henle, and finally the distal convoluted tubule which empties into a collecting duct.

The ureters are two muscular tubes approximately $25\ cm$ to $30\ cm$ in length that extend from the renal pelvis of each kidney to the bladder. They transport urine from the kidneys to the bladder through wave-like muscular movements called peristaltic contractions.

The bladder is a hollow, muscular, and elastic organ located in the pelvic cavity. Its primary function is to store urine until it is ready to be expelled from the body. The bladder wall contains specialized smooth muscle called the *detrusor muscle,* which contracts during urination. The average adult bladder can hold approximately $300\ mL$ to $500\ mL$ of urine. The urethra is the tube that ultimately carries the urine from the bladder out of the body during urination.

1.3.7 Immune

It is the purpose of the immune system to seek out and destroy harmful invaders such as bacteria, viruses, fungi, and other harmful or foreign substances, including cancer cells. It also plays a crucial role in identifying and eliminating abnormal cells, like cancer cells, and in clearing dead or damaged cells. Without a functioning immune system, our bodies would be constantly overwhelmed by pathogens, leading to severe illness and even death. To accomplish this mission, the immune system maintains an army of white blood cells which take the battle to the enemies by way of the circulatory and lymphatic systems.

Should an invader bypass the first line of defense, namely the skin, white blood cells such as neutrophils, eosinophils, basophils, or lymphocytes will descend upon the invader, engulf, and destroy the bacteria or virus. White blood cells such as monocytes can identify and neutralize dead cells, dead invaders, and bacteria and facilitate their removal. Cells such as *T* cells and *NK* cells can target and destroy cancer cells. White blood cells can also form antibodies which target and destroy known pathogens, forming what is called *natural immunity.*

In addition to their seek-and-destroy mission, white cells can also heal injuries by regulating the body's inflammatory response. They also help to regulate the immune system itself and prevent the targeting of healthy tissue or overreacting to a situation.

The immune system can be broadly divided into two main branches: the innate immune system and the adaptive (or acquired) immune system. These two systems work in concert, with the innate system providing immediate, non-specific defense, and the adaptive system offering a highly specific, long-lasting response.

Physical and chemical barriers form the very first lines of defense, preventing pathogens from entering the body in the first place. The skin is a tough and impermeable barrier that physically blocks most microbes. Mucous membranes line the inner surfaces of the respiratory, digestive, and urogenital tracts, producing sticky mucus that traps pathogens, which can then be expelled through coughing or sneezing. Cilia are hair-like projections in the respiratory tract that sweep mucus and trapped particles upwards where they can be expelled. The stomach provides a highly acidic environment that kills most ingested microbes. Tears and saliva contain enzymes like lysozyme that break down bacterial cell walls. Phagocytic cells actually "eat" or engulf foreign particles and cellular debris. Macrophages are large phagocytic cells found in tissues throughout the body. They are crucial for initiating immune responses and presenting antigens to adaptive immune cells. Neutrophils are the most abundant type of white blood cell and typically serve as first responders to any infection and are highly effective at engulfing and destroying bacteria; however, they have a short lifespan and contribute to the formation of pus. Dendritic cells are also phagocytic; however, their primary role is to capture antigens and present them to *T* cells. Natural Killer (*NK*) cells are lymphocytes that specialize in recognizing and killing virus-infected cells and some cancer cells. They do so by detecting a lack of "self" markers on the cell surface.

Inflammation is a localized tissue response to injury or infection characterized by redness, heat, swelling, and pain. This process attracts immune cells and other molecules to the site of infection/injury which helps to contain the infection and initiate tissue repair. The process is mediated by chemicals that are released by damaged cells: histamines, prostaglandins, and so on.

A fever is characterized by an increase in body temperature often triggered by pyrogens released during infection which attempt to inhibit the growth of some pathogens and enhance the body's immune cell activity.

The immune system can actually "learn" to recognize specific pathogens and develop a memory of them, thereby allowing for a faster, stronger, and more targeted response to any subsequent encounters. This is why we typically don't get the same infectious disease twice. This learning ability is largely due to the behavior of T and B cells. B cells are responsible for humoral immunity and the production of antibodies. Each B cell has unique receptors on its surface that can bind to a specific antigen. When a B cell encounters its specific antigen, it becomes activated. For most antigens, a B cell needs help from a T cell to become fully activated. The B cell processes the antigen and presents it on its surface. The T cell recognizes this presented antigen and provides cytokines that stimulate the B cell. Once activated, the B cell undergoes rapid proliferation, creating many identical copies of itself. These cloned B cells differentiate into plasma cells, which produce antibodies, or memory B cells, which persist in the body for years, sometimes decades, and can quickly differentiate back into plasma cells when needed, leading to a much faster and stronger secondary immune response.

1.4 The Essential Systems

While the aforementioned organ systems and our body's natural abilities and perception mechanisms might be very important to us, and perhaps critical for normal operation, they are not necessarily essential for the continued operation or sustenance of the body machine, as critical as they might seem. Many people are alive today with weakened or destroyed immune systems. Many people are alive today with faulty endocrine systems. There are people alive with skin cancer, bone diseases, and so on. These systems are important, and life with any of these being compromised might be hard, but life continues nevertheless. Only those functions which provide or deliver the requisite oxygen and nutrients to vital bodily organs are necessary to sustain the machine and therefore our lives, so if we boil that down to its core, there are only four: respiration, circulation, digestion, and the nervous system. If you consider the situation of complete life support, you will find that it includes these four and only these four.

1.4.1 Respiration

The function of the respiratory system can be summarized as a "simple" exchange process: provide oxygen to the blood while removing carbon dioxide and other waste gases. Like most vital systems, its function can be described quite succinctly, but the logistical operations are much more complex. In order for the respiratory system to perform its basic function, it must move several liters of air in and out of the body every minute while facilitating the bidirectional exchange of gases. Most of the respiratory components are dedicated to the physical movement of sufficient quantities of air, whereas only a single system handles the gas exchange process. This lopsided distribution of resources is largely due to air being a compressible fluid, which makes it relatively difficult to control as compared to something like a ball or potato. You cannot physically grab a pocket of air and force it into a balloon like so much dirt. You must induce the air to move by creating pressure imbalances, and that is a much more complicated and sometimes delicate process. Furthermore, the air might contain solids or liquids in addition to oxygen and nitrogen. These foreign elements must be removed before reaching the smaller regions where the gas exchange takes place. These items are large compared to the pathways the air will take, so these foreign entities would clog up the works and possibly cause physical damage to the respiratory system. Excessive amounts of water or other liquids can cause pneumonia and other diseases. Consequently, a number of filtration and conditioning systems are employed to sequentially remove any foreign elements and properly humidify the air to appropriate levels before they can cause harm.

1.4.2 Digestion

Eating and drinking are an essential part of life. The body requires fuel to operate and an assortment of building materials. A younger body may require these materials to produce or expand additional bone and muscle as the child grows; however, even an older body is in a constant state of maintenance, removing and replacing damaged or worn-out components. Skin is the number one item most people think about when considering renewal and replacement. On average, the outer layer of human skin only lasts about a month. Younger humans replace it faster, older humans slower. Infants less than a year old, for example, replace their skin every two weeks. Humans over the age of 60 may take up to three months to replace their skin entirely. Considering that skin is the largest of all human organs, that is a lot of building material required on an ongoing basis. However, there are a number of other cells that have a finite shelf life. The red blood cells, for example, are replaced every three to four months in an ongoing process of destruction, removal, and replacement. All of these building materials must be provided by the food and drink we consume on a daily basis, and it is the function of the digestive system to extract the needed and useful components from ingested foods, insert these materials into the bloodstream for delivery, and dispose of the rest along with extractions from the kidneys and spleen.

1.4.3 Circulation

The simplest model for the circulatory system is a giant conveyor system constantly carrying materials out and bringing materials back in a never-ending loop. The motor driving this conveyor system is the heart, a complex multi-stage hydraulic pump moving more than seven thousand liters of fluid per day on average through an intricate network spanning more than sixty thousand miles in combined length. Along the way, the blood picks up oxygen, carbon dioxide, various types of sugars and proteins, vitamins, and other minerals. The blood becomes a mega-market of supplies, which it carries out to the different cells and organs and muscles of the body where it partakes in a biological barter, exchanging needed oxygen and other materials for carbon dioxide and other waste products. On its way back, the blood runs through several filters, including the kidneys, liver, and spleen, which filter out both liquids and solids for disposal. Waste gases are exchanged in the lungs for a fresh supply of oxygen, and the cycle continues. Some excess nutrients, such as a variety of sugars, are sent into "warehousing," where they are put into storage as fat and can be called upon later, should the need arise. Perhaps this model is a bit simplistic and therefore lacks some degree of intricate precision, but it does accurately describe the *function* of the circulatory system if not the *details* of the operation.

1.4.4   The Nervous System

At the heart of the entire body, and the one thing keeping it all going, is the nervous system. Although the heart may (or may not) continue beating for some time without external stimulus, without a functioning nervous system, all of the muscles throughout the body would completely relax, including muscles in the chest, abdomen, arteries, and veins. A typical human has a total blood volume of about 5 liters; however, if all of the veins and arteries were to suddenly relax, the combined volume would be more than 40 liters. While the nervous system is operating, the veins and arteries are in various stages of contraction. This is one of the methods by which the nervous system regulates blood pressure. Other methods involve regulating the operation of the heart, telling it how fast and how hard to beat. The nervous system also regulates the muscles required for breathing in and out, constantly monitoring the levels of oxygen and carbon dioxide in the bloodstream as well as heart rate, blood pressure, body temperature, and so on. All of these command and control operations are located in the brain stem, more specifically the portion known as the Medulla Oblongata. This portion is completely autonomous, which is why you do not have to consciously think about breathing or making your heart beat. The other parts of the brain stem, the pons, and midbrain, each have some degree of correlation with the higher brain and consequently have more conscious interaction. The pons, for example, helps with balance and equilibrium, whereas the midbrain works with the eyes and ears. Both of these share significant communications with the cerebrum and cerebellum, regions with more conscious control and interaction.

1.4.5 Four Corners of One Square Human

As this outline details, respiration, digestion, circulation, and the nervous system each form the four corners of mutual support in human survival, the four key systems that are all equally essential to support a single living human. Without even one of these systems, the human cannot survive, and life ends. It may end in minutes, hours, or days, but the body will no longer be able to sustain itself and ultimately stop functioning.

Without the addition of oxygen by the respiratory system, the cells would stop functioning within a couple minutes. Cellular collapse resulting in permanent cell termination would follow a few minutes later. Without the ability to shed carbon dioxide, the body would rapidly lose its ability to function, the blood would lose its capacity to carry oxygen, and the nervous system would start to malfunction, resulting in seizures or loss of consciousness, and ultimately the entire body would cease to function.

Without digestion the body would run out of fuel. Cells require water, sugars, and other nutrients in order to operate properly, as well as proteins and vitamins for growth and repairs. These materials are provided by the foods that are digested. The body can call upon the fuel stored in the fatty tissues for a while, but the rate at which these sugars can be extracted and used by the cells is limited and generally less efficient than regular consumption and digestion. Furthermore, the other materials required by the cells in order to survive, reproduce, and repair themselves are not necessarily available. Nevertheless, eventually all available fuel will be exhausted, and all remaining processes will come to a stop, and the body will die. This could take days or weeks, but the outcome is inevitable.

Without circulation, there is no mechanism available for the oxygen, water, sugars, proteins, or anything else to reach the cells and organs of the body. In this case, the body stops functioning almost immediately.

The autonomic nervous system is necessary to regulate and control all of these processes as the central governing authority, hence the name *central nervous system,* and without it most of these operations would not be carried out. The only operation with limited autonomy is the heart, but without oxygen or a fuel supply, even that would not last more than a few minutes.

The primary focus of this text is on the respiratory system, but understanding the mutual interdependence of each key life system is important in order to understand and appreciate the complexity that is the human life form. The body truly is the most complex machine in existence, the likes of which have yet to even be approximated by anything artificially created.

2

Compressible Dynamics of Respiration

2.1 Classical Review

Before embarking on a rigorous study of fluid dynamics and human anatomy, we should take a moment and get our concepts sorted by conducting a brief review of classical mechanics, some definitions that we will find useful, and some basic concepts that will be referenced frequently as we move forward. It is important to have a common understanding of each of these key concepts before moving forward.

2.1.1 Vectors versus Scalars

Most people are quite familiar with numbers, even if they really despise them. Ask a person how much an apple costs, and they will give you a number. Ask how many players are on the field, and they give you a number. It is natural. We have been counting things and quantifying things since we were young children. It was probably the second thing you learned to do after talking.

Numbers such as 5 *apples* and 11 *players* are magnitude quantities that represent how much of something you have. It consists of two parts: the *magnitude* or size in the form of a number, and the label or *unit* which indicates to what the number refers. For example, the magnitude is 5, meaning you have five of them, and the label is *apples*, meaning the magnitude is referring to how many *apples* specifically as opposed to any other thing. These quantities are known as *scalars* and are typically positive (since the minus sign belongs to the direction). After all, you cannot have "negative five" apples.

Other quantities require one more piece of information. If you are giving a tourist directions on how to get to a certain location, you cannot simply tell them to drive 25 *km*. You must also tell them "which

DOI: 10.1201/9781003683476-2

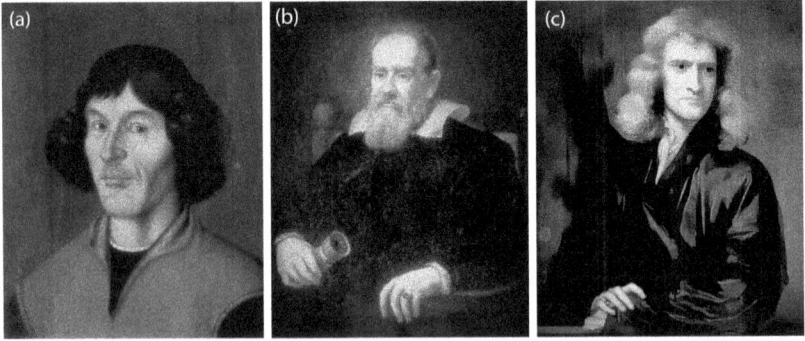

FIGURE 2.1
(a) Nicolaus Copernicus (1473–1543) from a portrait in the Museum in Toruń, (b) portrait of Galileo Galilei (1564–1642) by Justus Sustermans in 1636, and (c) Isaac Newton (1643–1727) from a portrait painted in 1689 by Godfrey Kneller. All images are in the Public Domain. These individuals collectively contributed to the foundations of all science. Copernicus is best known for his heliocentric solar system, Galileo for the concept of inertia and gravity, and Newton for his fundamental laws of motion.

way," meaning *North* or *East* or whatever. This is why we call it "giving directions," because you must include the *direction* in your instructions. You tell them to drive 25 *km East*, for example. "Get on this highway and drive *North* for 60 *km*." In addition to the magnitude and unit, you include a direction. These quantities are called *vectors*.

We differentiate a vector quantity by including an "arrow" over the symbol, such as $\vec{v}$. Examples of vector quantities include force $\vec{F}$, velocity $\vec{v}$, acceleration $\vec{a}$, area $\vec{\mathcal{A}}$, and momentum $\vec{p}$. Quantities that do not have such an arrow over them are considered to be scalars. Examples of scalar quantities include mass m, volume $\mathcal{V}$, time t, and pressure P.

2.1.2 Newton's Laws

Building off the work of Nicolaus Copernicus (1473–1543) and Galileo Galilei (1564–1642), Isaac Newton (1643–1727) also shown in Figure 2.1 consolidated our understanding of forces and motions into three laws that describe the relationship between the forces acting on an object and the resulting motion, or more specifically *change* in motion, of that object.

First Law: The Law of Inertia

In the first law, Newton stated:

> "The velocity of an object will remain constant unless acted upon by an external net force."

This was revolutionary at the time, since Aristotle claimed all objects would ultimately return to rest (stationary). There are several subtle points about this first law that might escape notice at first glance.

Notice that Newton carefully uses the term "velocity" and states that the velocity will remain constant. Velocity is a vector quantity possessing both magnitude and direction. For a vector to remain constant, both the magnitude and direction must both remain constant. That means the object cannot speed up, slow down, or even turn.

Second point has to do with the "external net force." Notice that Newton carefully puts constraints on the type of force that is being indicated. The force must be external to the object. As any astronaut can attest, you cannot exert a force on yourself. You must push on the wall of the spacecraft, and Newton indicates why in this first law. The wall is external to the astronaut.

The other constraint is the use of the term "net force." Newton did not say *any* force, he said *net* force. That means total sum of all forces acting on the object. Since force is a vector quantity, that means you have to add up the forces in each of the three cardinal directions: up/down, left/right, front/back. The velocity of the object will be constant unless at least one of these is out of balance.

Second Law: The Law of Acceleration

The second law is the only one written almost exclusively in its mathematical form:

$$\vec{F}_{net} = m\,\vec{a} \qquad (2.1)$$

where $\vec{F}_{net}$ is the net force acting on the object, m is the mass of the object, and $\vec{a}$ is the resulting acceleration, or change of velocity, of the object.

Mathematics is the language of Physics, and it is a highly compressed and sometimes complex language. As simple as Equation 2.1 looks, this simple expression explains the orbits of the planets, the flight of a 747, it got us to the moon and back, it landed the Mars rover, and on and on and on. Three simple letters and one symbol did all that. This is a perfect example of the "beauty" of nature and just how much information can be compressed into a single and seemingly trivial mathematical expression. The key to mastering Physics is learning how to "decode" and understand this new language.

Equation 2.1 indicates that the net force is directly proportional to the acceleration of an object, and that constant of proportionality is the mass, m. That means a graph of the magnitude force and its resulting acceleration will be that of a line with slope m. Galileo first discovered this relation and used the term *inertia*. He described it as the resistance of an object to any applied force that was attempting to cause an acceleration. Newton later changed the term to *mass* as we know it today, but the concept of inertia remains valid. If you double the force on an object, you will double the acceleration. That is what *directly proportional* means. However, if you double the mass, the acceleration will be cut in half as the object has now doubled its ability to resist any change in its velocity.

Third Law: The Law of Reciprocity

The third law is often stated

"For every action there is an equal and
opposite reaction."

Many authors take on language involving *action-reaction pairs* and talking about different types of forces. The bottom line is quite simple, at least mathematically.

$$\vec{F}_{AonB} = -\,\vec{F}_{BonA} \tag{2.2}$$

In other words, the force of A on B is equal in magnitude to the force of B back on A. The minus sign belongs to the direction, so they are equal in magnitude and opposite in direction. In reference to our earlier example, the astronaut pushes on the wall; however, their acceleration is actually a result of the wall pushing back on them. Equation 2.2 indicates the wall pushes back on the astronaut as a result of the astronaut pushing against the wall, and Equation 2.1 indicates the direction of the resulting acceleration must be the same as the direction of the net external force, which is the wall pushing back on the astronaut. Hence, the astronaut moves away from the wall.

2.1.3 Kinematics

An object at rest shall remain at rest unless acted upon by a net external force, or so said Isaac Newton. In other words, the *position* of the object shall remain a constant. Kinematics is the study of the mathematics (matics) of motion (kine). It is a mathematical way to describe an object's position $\vec{r}$, velocity $\vec{v}$, and acceleration $\vec{a}$ over time t.

Every object in the universe has its own, unique position, since no two objects can occupy the same physical space. An object's velocity is defined as the change in that unique position over time, or

$$\vec{v} = \frac{\partial \vec{r}}{\partial t} \tag{2.3}$$

Similarly, the acceleration of an object is the change in an object's velocity over time.

$$\vec{a} = \frac{\partial \vec{v}}{\partial t} \tag{2.4}$$

We can also couple that with Equation 2.3 to get

$$\vec{a} = \frac{\partial}{\partial t}\left(\frac{\partial \vec{r}}{\partial t}\right) = \frac{\partial^2 \vec{r}}{\partial t^2} \tag{2.5}$$

If we start with Equation 2.4 and integrate over time, we get

$$\int \frac{\partial \vec{v}}{\partial t}\, dt = \int \vec{a}\, dt \tag{2.6}$$

which gives us our first kinematics equation

$$\vec{v}_f = \vec{v}_i + \vec{a}\, t \tag{2.7}$$

On the other hand, if we start with the second expression in Equation 2.5 and integrate over time, we get

$$\int \frac{\partial^2 \vec{r}}{\partial t^2} \, dt = \int \vec{a} \, dt \tag{2.8}$$

or

$$\frac{\partial \vec{r}}{\partial t} = \vec{v}_i + \vec{a} \, t \tag{2.9}$$

where we have used Equation 2.7 to resolve the first integration constant, $\vec{v}_i$. Integrating a second time, we get our second kinematics equation

$$\vec{r}_f = \vec{r}_i + \vec{v}_i \, t + \frac{1}{2} \vec{a} \, t^2 \tag{2.10}$$

These are two of the four kinematic expressions that describe the position and motion of an object as a function of time.

2.1.4   Uniform Circular Motion

Newton's Second Law which is expressed in Equation 2.1 states that the velocity of any object will remain constant, both in magnitude and direction, unless an external net force acts upon it; therefore, an object moving in a circular path must have an external net force causing it to do so. This force is the *centripetal force* which is given by

$$F_c = r \, \omega^2 \tag{2.11}$$

where r is the radius of the circular path and ω is the angular velocity of the motion. The linear velocity v of the object along this circular path is directly related to the radius r of orbit by

$$v = r \, \omega \tag{2.12}$$

so we can write Equation 2.11 as

$$F_c = \frac{v^2}{r} \tag{2.13}$$

The commonly used term *centrifugal force* is actually a consequence of the object's inertia, or mass. Objects want to travel in a straight line, and the centrifugal force is the sensation you get when you are forcing your body to do otherwise. Sediment will settle to the bottom of the test tube, not because of any centrifugal force, but because that is where it has to go in order for the glass at the bottom of the tube to be able to direct an appropriate magnitude and direction centripetal force on the sediment.

The term *centripetal* comes from a combination of the Latin terms "centrum" and "petere" and translates approximately to "center-seeking;" whereas, the term *centrifugal* comes from the Latin terms "centrum" and "fugere" and translate to "center-fleeing."

Both were originally used in Newton's *Philosophiae Naturalis Principia Mathematica,* or just Principia, and date back to 1687.

2.1.5 Hooke's Law and Harmonic Motion

Solids and liquids are held together by interatomic bonds which can vary in strength and type. It is also possible that two atoms might share more than one bond of a particular type forming what are called double and triple bonds. Each individual bond extends from one atom to another from center to center, but the arrangement of the atoms varies depending on the identities and quantities of each atom involved in the material. The chemical formula for a substance identifies the basic building block and the numbers of each constituent atom, such as H_2SO_4 or Fe_2O_3. These atoms are bonded together in the liquid or solid, respectively, that make up the substance. Determining and explaining the shape of these building blocks is the focus of the research field known as molecular geometry. These basic shapes are of significance since they determine the possible shapes of their larger samples and play a large role in the structural parameters such as hardness, flexibility, temperature resistance, conductivity, and so forth.

The bond between adjacent atoms in a solid or liquid can be approximated to high precision by using Hooke's Law which states the amount of force is directly proportional to the amount of extension or compression, as given by

$$\vec{F} = -k \, \vec{\Delta x} \tag{2.14}$$

where $\vec{F}$ is either the applied force or resistant force created by the spring, k is the *spring constant* in units of *Newton* per *meter*, and $\vec{\Delta x}$ is the modification in the length of the spring or bond. This relation is negative because Hooke's Law represents what is known as a *restoring force*, in that the opposition force posed by the spring will always seek to return the spring to its original length, known as the "natural" or "rest" length.

    If we track the motion of the object attached by a spring to a rigid point, the spring force, and thus the net force, on the object will be a maximum when the spring is extended or compressed to its maximum extent. This will be the points where the acceleration is also maximum and the mass will be momentarily at rest. As the mass passes the neutral point, where the spring does not exert a force because it has returned to its natural length, the object will continue to move in accordance with Newton's First Law. In the absence of any degrading forces, such as friction, the object will continue to oscillate between maximum extension and maximum compression repeatedly and indefinitely in what is called *simple harmonic motion.*

    If we assume Equation 2.14 is the one-dimensional net force on an object of mass m and apply Newton's Second Law in Equation 2.1, we get

$$m \, \frac{\partial^2 x}{\partial t^2} = -kx \tag{2.15}$$

where we have used Equation 2.5 as our acceleration. Rearranging Equation 2.15 and isolating our x terms we get the differential equation:

$$\frac{\frac{\partial^2 x}{\partial t^2}}{x} = -\frac{k}{m} = -\omega^2 \tag{2.16}$$

where we have introduced

$$\omega = \sqrt{\frac{k}{m}} \tag{2.17}$$

The solution to this differential equation is as follows

$$A(t) = A_0 \, Sin(\omega t) \tag{2.18}$$

where $A(t)$ is the amplitude of the oscillation at some time t, A_0 is the initial amplitude, and ω is given by Equation 2.17. Since the sine function ranges from -1 to $+1$, the "amplitude" is defined as the distance the object has been displaced from the natural length of the spring, known as the neutral point where the net force is 0. The locations of maximum amplitude, either $+A_0$ or $-A_0$, are called *turning points* since the mass is literally in the process of reversing direction. If we map out the amplitude over time, we will clearly get a *sine wave* since Equation 2.18 is literally the sine function. A_0 would then be called the *amplitude of the wave*. The term ω is known as the angular velocity and is related to the frequency of oscillation by

$$\omega = 2\pi f \tag{2.19}$$

where f is the commonly understood frequency in units of Hz or *cycles per second*. This literally refers to how many times the mass makes one complete round trip and returns to its original location every second.

2.1.6 Pressures, Forces, and Areas

Pressure is defined as a force F acting upon an area $\mathcal{A}$, or

$$P = \frac{F}{\mathcal{A}} \tag{2.20}$$

and has units of *Pascals* (Pa), in honor of Blaise Pascal (1623–1662) and is given by

$$1.0 \, \frac{N}{m^2} = 1.0 \, Pa \tag{2.21}$$

One may notice that, unlike force, pressure is not a directional quantity. Indeed, it is *omnidirectional* and can act in all directions simultaneously. For that reason, pressure is a *scalar,* not a *vector,* in spite of the fact that both force *and* area are both vector quantities.[1]

---

[1]It is interesting to note that, while the multiplication of vectors can take four different forms, division of one vector by another vector is not defined, and its presence in any formulaic process is typically indicative of some mathematical error.

There are two kinds of pressure: internal and external. Internal pressures are those literally inside a substance. While these are most often associated with gases, solids and liquids can also have internal pressures when placed under the influence of external forces. Water, for example, has an internal pressure that is directly proportional to its depth. Even large steel beams can compress like a spring under the weight of fifty floors of an office building. As these examples indicate, both solids and liquids have natural volumes. Consistent with Newton's Third Law, any attempt to compress them will result in a higher internal pressure to oppose that change.

Unlike solids or liquids, gases do not have fixed or preferred volumes. A rock will sit at the bottom of a glass and remain a rock with the same shape and volume it had moments before it was placed in the glass. Water or any other liquid will take the conform to the shape of the glass but reside in the bottom with a fixed and definite top surface. Its shape might vary as it conforms to the container, but its volume remains fixed. A gas, on the other hand, will expand and occupy the entire volume of the container. If placed in a conventional glass, it will escape the glass and expand to fill the entire room. This is why you can smell perfume all the way across the room and everywhere within the room. The gas particles from the perfume slowly migrate their way until they are equally represented everywhere in the room.

Applying a force on a brick or stick is all well and fine, but you cannot apply a force on a puddle of water. The water simply moves out of the way as your hand goes by. If you wish to make the water in the puddle up and move, gain and sustain a velocity, you must apply a *pressure,* and this is true for any *fluid,* generally. We will formally define the term "fluid" later in this section, but both liquids and gases are always a fluid and require pressure, not just force, to cause them to move in any meaningful way.

2.1.7   Stress and Strain

When applying a force to a surface area, there are two options: the force can be perpendicular to the surface, or it can be parallel to the surface. The result is a change in the bond lengths or bond angles, depending on the direction of the force.

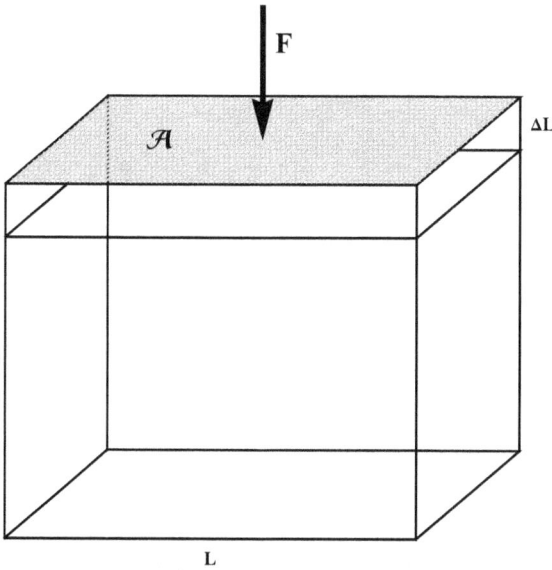

FIGURE 2.2

In the event the force acts perpendicular to the surface area, the height of the cube will be decreased by an amount known as the strain. Consequently, the volume of the enclosed space will also decreased by an amount proportional to the amount of strain thereby increasing the density of the material.

2.1.7.1 Normal Stress

If the applied force is perpendicular to the surface area, as shown in Figure 2.2, the material is put under stress. Since the force is perpendicular to the surface, this is known as a ***normal force*** $F_{\perp}$ and therefore this produces a ***normal stress*** on the material

$$\sigma_N = \frac{F_{\perp}}{\mathcal{A}} \tag{2.22}$$

where σ_N is the normal or tensile stress and $\mathcal{A}$ is the surface area being acted upon. As a result of this stress, the material will suffer some amount of *strain* ε_N, or change in its characteristic length parallel to the applied perpendicular force $F_{\perp}$, as indicated in the diagram. The actual amount of linear change is proportional to the initial length, as the stress will be distributed along all of the molecular bonds, so

$$\varepsilon_N = \frac{\Delta L}{L} \qquad (2.23)$$

This results is an overall change in the volume of the material according to

$$\Delta V = L^3 \, \varepsilon_N \qquad (2.24)$$

Since the mass of the material remains constant but must now reside in a new volume, this results in a change of density as given by

$$\Delta \rho = \frac{m}{\Delta V} = \frac{m}{L^3 \, \varepsilon} \qquad (2.25)$$

Notice that the change volume V is in the denominator; therefore, if the volume *decreases* the density will correspondingly *increase*.

The ability for a substance to change density under a given amount of stress is known as the *Bulk Modulus* and is given by

$$\partial P = -B \frac{\partial V}{V} \qquad (2.26)$$

which gives

$$B = -V \frac{\partial P}{\partial V} \qquad (2.27)$$

which clearly shows the change in volume with pressure. Furthermore, the negative slope indicates a *decrease* in volume with *increasing* pressure, as expected. Water is notably *incompressible,* as are all liquids, and has a Bulk Modulus of 2150 *MPa*; whereas air, which is highly compressible as are all gases, has a Bulk Modulus of 101 *kPa*, more than four orders of magnitude smaller than water.

2.1.7.2 Shear Stress

If the applied force is parallel to the surface area, as shown in Figure 2.3, the material is put under a different kind of stress. Since the force is now parallel to the surface, it attempts the *shear* the upper surface with respect to the lower surface. This is known as a **shear force** $F_\parallel$ and it produces a **shear stress** on the material

$$\sigma_s = \frac{F_\parallel}{A} \qquad (2.28)$$

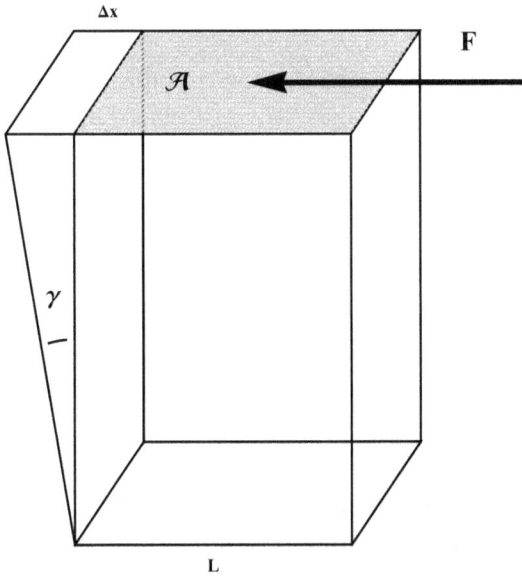

FIGURE 2.3
In the event the force acts parallel to the surface area, the cube will be deformed.

where σ_s is the shear stress and $\mathcal{A}$ is the surface area being acted upon. As a result of this stress, the material will suffer some amount of deformation as the upper side moves an amount Δx with respect to the lower fixed side. The *strain* ε_s is defined as the ratio of the two sides of the triangle as a result of the applied shear force $F_\parallel$, as indicated. Thus

$$\varepsilon_s = \frac{\Delta x}{L} \tag{2.29}$$

The small angle γ indicated in Figure 2.3 can be determined as follows

$$Tan\gamma = \varepsilon_s = \frac{\Delta x}{L} \tag{2.30}$$

and is also often referred to as the shear strain. This is not entirely in contradiction, since the change Δx is often much smaller than height L, thus $\gamma \approx \varepsilon_s$.

2.1.7.3 Young and Yield

The ratio of stress to strain is known as the *Young's Modulus*

$$Y = \frac{\sigma}{\varepsilon} \qquad (2.31)$$

The Young Modulus is defined separately for tensile (normal) and shear conditions. The postulate suggests that the amount of shear is directly proportional to the applied stress, so the value of Y for a given material is a constant so long as the stress applied does not exceed what is known as the *elastic limit* for the material. At this point, the material is no longer able to return to its original dimensions or condition. It becomes permanently deformed in the respective direction. If the stress is too great, the material breaks. This stress is called the *yield point* for the material. In the case of normal or tensile stress, the rope breaks. In the case of shear stress, the material is severed, which is why scissors are also known as shears.

2.1.7.4 Crystals and Fluids

The applications of normal and shear stress work well for solid material because of their strong molecular bonds holding the material together. A solid has a single, fixed shape which, at the molecular level, is built up of small shapes similar to fractals that govern the limitations on what the larger structure can become. For example, a diamond will never be round no matter how hard you try because its basic building blocks are triangles. This strong bond between their elementary building blocks allows for some elastic behavior for a solid, and the fact they are built up out of these small fractals is why all materials of this type are called *crystals,* not just diamonds and rock salt. At the molecular level, all solids are crystalline in nature, only the shape of those fundamental crystals changes from one material to another.

Life gets complicated when you deal with non-crystalline materials. Such compounds lack the strong cohesion of a crystal and often lack any fundamental structure. Collectively these materials are called *fluids,* and, in spite of their numerous individual differences, they all share certain traits in common that qualify classifying them all in this manner.

First and foremost, the lack of any strong molecular cohesion means they lack the ability to oppose any sort of shear stress in the same manner that a crystal would. If you put a shear stress on a fluid, the deformation will happen continuously resulting in a *flow*. There may be some opposition to that flow in the form of a *viscosity*, but the flow will happen nonetheless. If you stop the shear stress, most fluids do not rebound, and none of them return to their original shape. In fact, such materials are called *amorphous* because, unlike crystals, they do not *have* a particular shape. They can take any form required for any container. Liquids will group together at the bottom of any container, such as a glass or pond. Gases expand to occupy the entire amount of volume available, be it inside a sealed container or surrounding a planet. The atmosphere is spherical because the Earth is spherical, the water in the glass conforms to the shape of the glass, and so on.

Normal stress is a not so simple because it is generally not possible to "pull" on a fluid. We can, however, "push" quite effectively and that leads to the issue of compressibility indicated in Equation 2.26. Liquids are typically taken to be *incompressible*, meaning they do not change volume or density no matter how much pressure is applied. This is not in contradiction to Equation 2.26, as this concept is widely used in the field of industrial hydraulics where pressures in excess of 5000 *psi* are common. However, gases are highly compressible which means they change their volume and density quite easily. Unlike most liquids, gases are relatively immune to significant complications due to high temperatures, so this makes this property quite useful in the field of pneumatics.

2.1.8 Waves, Lengths, and Frequencies

The elastic nature of materials as described by Equation 2.18, including both crystals and fluids, allow disturbances to travel through them at a speed characteristic to the strength of their molecular bonds or bulk modulus. These traveling disturbances are what we know as waves. The size of the individual disturbance is known as the *amplitude* $A(t)$, and the number of disturbances that pass per second is known as the *frequency* f. As the disturbance moves through some material, the spatial separation between adjacent crests is known as the *wavelength* λ. The speed of travel v through the material is related to this quantities by the following

$$v = f \lambda \tag{2.32}$$

where λ is the wavelength, f is the frequency, and v is the wave speed.

Although wavelength and frequency appear in Equation 2.32 to be related to the wave speed, the actual speed of the wave through the material does not depend on the frequency nor the wavelength. The speed of a wave through any material, be it air, water, or solids, is strictly dependent only on the physical properties of that material as indicated by Equation 2.18 since ω is defined by Equation 2.17. As a general rule of thumb, harder and more rigid materials have larger spring constants k and thus tend to have higher wave speeds; whereas, softer and more pliable materials have smaller k values and lower speeds. For an example, steel is incredibly hard and has wave speeds of approximately $5000\ m/s$. Water is very soft and pliable and has wave speeds of about $1000\ m/s$ depending upon temperature and salinity. Air is very light and has virtually no tensile strength whatsoever; consequently, the speed of sound is only about $300\ m/s$ depending on temperature and humidity.

2.1.9   Fahrenheit and Celsius

Understanding the behavior of gases first requires an overview of some principles of general kinematics and statistical thermodynamics, starting with the definition of temperature. General kinematics such as those covered in Section 2.1.3 deals with the laws of motion and are related to kinetic energy, or the "energy of motion." Statistical thermodynamics directly relates this "energy of motion" to the temperature of a gas, which in turn is related to the pressures within that gas. The movement of a fluid is governed by pressures not forces, and that includes the movement of both liquids and gases. Since pressure is directly linked to velocities, as covered in Section 2.1.6, and pressure governs the motion of fluids, to understand the motion of any fluid we must first deal with what we mean by temperature.

Most people think of temperature in terms of Fahrenheit or Celsius. If you ask most people what a "comfortable" room temperature might be, they will answer either $68\ ^\circ F$ or $20\ ^\circ C$. These numbers, and in fact all such numbers using the term *degrees,* are arbitrary units on equally

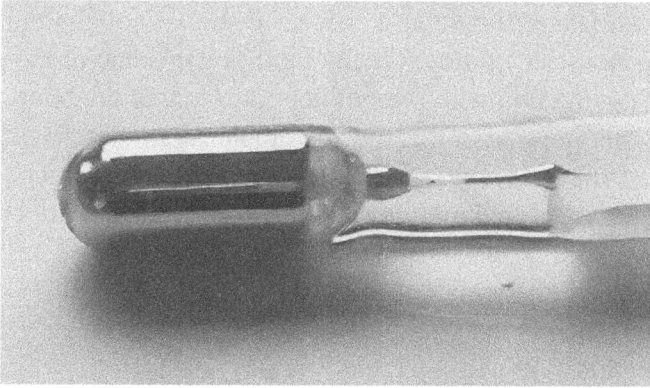

FIGURE 2.4

The mercury reservoir of a thermometer is shown here. You can also see the small capillary tube leading up the shaft. The small constriction in the capillary tube suggests this is a medical thermometer designed for easy upward flow but resistant to downward flow. This is why people had to "shake" the thermometer before use. This action would drive the mercury back into the reservoir. Heating would then cause the small column to rise again.

arbitrary scales. Neither of them are based on first-principles physical behavior, so neither have any physical meaning.

In 1714, Daniel G. Fahrenheit (1686–1736) constructed the first standardized and reliable thermometer. The design concept involved the expansion of a fluid in a uniform chamber. Mercury was chosen due to its high boiling point and very high surface tensions which prevent wetting of the interior surfaces. Mercury is very strongly attached to itself and is generally phobic to any other materials, which is why it forms into balls when dropped onto a table or other surface. The design of the chamber was of critical importance, since liquids expand volumetrically. Solids will expand their linear dimensions because of their linear bonds, but fluids expand dynamically in all directions. While Fahrenheit did not understand the reasons behind it, this behavior was in fact well known.

The chamber chosen includes two parts: the bulb and the shaft as shown in Figure 2.4. In the bulb, a relatively large quantity of the thermal substance, mercury, was stored. An extremely small diameter tube extends from the bulb up a calibrated shaft. Enough fluid is placed in the thermometer in order for the fluid to rise some distance, and to allow for contraction when the fluid cools beyond the temperature of construction. The size of this tube and the uniform diameters were the hardest part for this early 18th century scientist. Fahrenheit had to blow his own glass and make his own thermometers. Once he had finally achieved the skills and techniques to create repeatable and reliable instruments, all that remained was calibration.

A liquid such as mercury will expand its volume linearly according to the equation

$$\Delta V = \beta V_0 \Delta T \tag{2.33}$$

where ΔV is the change in volume, V_0 is the initial volume at the initial reference temperature, ΔT is the change in temperature from the reference point, and β is the volume expansion coefficient of the liquid in question. Since the expansion tube is cylindrical in shape, this change in volume can be registered as

$$\Delta V = \pi r^2 \Delta h \tag{2.34}$$

where r is the radius of the cylinder and Δh is the change in height of the column of fluid. The assumption here is that the expansion of the radius of the cylindrical chamber is negligible to the change in height, or at least they would track together in some appropriate manner providing a nice, linear relationship. This assumption works out well, and the thermometers were eventually a success. Putting Equations 2.33 and 2.34 together, we arrive at the linear expression

$$\Delta h = \frac{\beta V_0}{\pi r^2} \Delta T \tag{2.35}$$

All that was necessary was to determine the numbers on the calibrated scale. To achieve the necessary calibration, Fahrenheit needed two points to use as reference, then he would simply compute the slope and distribute the numbers accordingly.

Fahrenheit chose the numbers 0 and 96. Choosing 0 seems an obvious choice and one that Celsius would later do as well. Choosing 96 is a little less obvious, but this number has the remarkable characteristic of being evenly divisible by a relatively large number of integers, and decimal values were not widely used in 1714, so it seemed an elegant choice at the time. For 0, Fahrenheit chose the equilibrium temperature of a brine solution made of 1/3 water, 1/3 salt, and 1/3 ice. This solution got extremely cold. He chose body temperature to be 96; however, he used his armpit as the reference measure. As we know today, this is actually the *axillary* temperature, not the core body temperature. As a result, the true internal body temperature is known today as $98.6°F$. Chopping the two lines on his new thermometers into 96 equal parts, we find fresh water freezes at $32°F$ and boils at $212°F$.

Several decades later, and toward the end of his career and his life, Anders Celsius (1701–1744) developed a new scale in 1742 divided into an even 100 parts starting with the freezing point of fresh water at $0°C$. His reasoning was the invariability of this point when changing latitude or pressure making this point reliable. Citing the previous works of Fahrenheit regarding the variability of boiling point with pressure, Celsius further stipulated $100°C$ would be the boiling point of fresh water at an atmospheric pressure of 25.3 *in Hg*, which is approximately 0.846 *atm* of pressure. Standard atmospheric pressure is, of course, 1 *atm* or 760 *mm Hg* or 29.9 *in Hg*.

2.1.10 Boltzmann and Kelvin

While Celsius tried to be more physically reliable, neither temperature scale had any connection to physical laws. Both relied upon the observed behavior of a chosen substance, even if under specific circumstances. To make a physical connection, we must look deeper into *why* the substance, water in this case, changes from one form to another. *Why* does the liquid, mercury for example, expand when heated? The answer lay in the true meaning of temperature.

In 1848, more than a century after Celsius, Lord Kelvin, born William T. Kelvin (1824–1907) shown on the right in Figure 2.5, proposed an absolute temperature scale. On this scale, all values would be necessarily positive. Originally, this scale would coincide with the *triple point* of water, a curious point where the solid, liquid, and gaseous forms of water coexist in the same mixture at the same time. The chosen point was to be 273.15 on the Kelvin temperature scale. In the later part of the 19th century, Ludwig Boltzmann (1844–1906) shown on the left in Figure 2.5 looked into the statistical behavior of large numbers

(a)

(b)

FIGURE 2.5
(a) A photograph of Ludwig Boltzmann (1844–1906) taken in 1902. (b) a photograph of Sir William Thomson, Baron Kelvin (1824–1907) originally taken around 1900 by T. & R. Annan & Sons and restored by Adam Cuerden. All pictures and restorations are in the Public Domain.

of molecules, laying the foundation for what would become statistical thermodynamics, a.k.a. statistical mechanics. Boltzmann expressed the connection between the statistical behavior of the molecules and the temperature of the gas, but it was not until 1900 when Max Planck (1858–1947) derived and quantified the relationship:

$$\frac{1}{2}m\,v^2 = \frac{f}{2}\,k_B\,T \tag{2.36}$$

where m is the mass of the molecule, v is the *mean-square* speed of the molecule, T is the absolute temperature in *Kelvin*, f is the *degrees of freedom*, and k_B is the Boltzmann Constant given as

$$k_B = 1.380\,650\,524 \times 10^{-23}\,\frac{J}{K} \tag{2.37}$$

For the first time, we have a temperature scale that is directly linked back to first-principle concepts and behavior. For a mono-atomic gas, where the gas is made up of single atoms like so many billiard balls, $f = 3$ which represents the three cardinal directions: x, y, and z. For diatomic molecules, such as O_2 or N_2, $f = 5$ in order to include rotational motion. For all other molecules, such as CO_2, the degrees of freedom can include vibrations and other oscillatory modes with values of 7 (triatomic) or even 9 (polyatomic). If we assume the gas is mono-atomic with $f = 3$, then we can deduce the mean-square speed of our molecules to be

$$v_{rms} = \sqrt{\frac{3\,k_B\,T}{m}} \tag{2.38}$$

If we use Nitrogen ($m_{N_2} = 28.0134\ u$), which accounts for nearly 80% of Earth's atmosphere, with standard $20°C$ temperature and 1 atm of pressure, we get an average translational speed of

$$v_{rms} = \sqrt{\frac{3\,\left(1.380\,650\,524 \times 10^{-23}\,\frac{J}{K}\right)(273.15\ K)}{28.0134\,(1.665\,402 \times 10^{-27}\ kg)}} = 492.45\ m/s \tag{2.39}$$

where we have used the fact that $1\ u = 1.665402 \times 10^{-27}\ kg$. It is important to note that this is the mean-square speed. Not all of the molecules are going this fast, in fact probably very few are going this exact speed. Some are going faster, others slower, but *on average* this is a good estimate. Furthermore, this mean-square speed is independent of direction. A component of this derivation is the assumption that all directions (degrees of freedom) are equally possible.

2.1.11 Momentum and Pressure

Pressure is generated through collisions, either with other gas molecules or the bounding walls of some container. It is something of a misconception to actually think of a gas molecule or atom actually "colliding" with another particle or container, but the difference is more conceptual than practical; however, the realization culminates in truly ideal collisions that conserve both energy and momentum, and that is something that does not happen at scales familiar with most humans. Even the balls on a billiard table only approximate ideal to within 90%. But, at the molecular scale, all collisions are perfect meaning neither energy nor momentum is lost in the process. This is how a confined gas can retain its constant temperature and pressure indefinitely. It also plays a role in how energy is distributed, or diffused, throughout the gas ultimately resulting in all gas molecules sharing roughly the same amount of energy, on average.

Momentum is defined as the product of mass and velocity, or

$$p = mv \tag{2.40}$$

When a gas particle collides with the container wall, the part of the velocity perpendicular to the container wall changes direction, but the magnitude speed stays the same due to energy conservation. If we assume the gas molecule hits the wall head-on, perpendicular to the wall, then

$$\Delta p = m\Delta v = 2\,mv = 2\,p \tag{2.41}$$

so the magnitude change in the momentum is actually twice that of the original momentum. Furthermore, the net change in momentum for the system, including both gas and container, must be 0. After all, the bottle of air is not likely to jump off the table on its own. That means the container wall must also suffer this same change in its momentum.

The change in momentum is defined as an *Impulse*, and that change in momentum over time equates to a force.

$$F = \frac{\Delta p}{t} \tag{2.42}$$

If we average the number of collisions between the gas molecules and the corresponding container wall over time, we can determine the time-averaged impulse force on the chosen container wall.

$$<F> = \left(\frac{N_t}{t}\right)\Delta p \tag{2.43}$$

where N_t is the time-averaged number of collisions per second in our chosen direction. Since force over area is the definition of pressure, we can divide this time-averaged force by the corresponding surface area of the container wall and get the time-averaged pressure of the gas on the respective container wall.

$$P = \frac{<F>}{\mathcal{A}} = \frac{N_t\,\Delta p}{\mathcal{A}\,t} \tag{2.44}$$

Since the change in momentum depends on speed, we get

$$P = \frac{2\,N_t\,m\,v}{\mathcal{A}\,t} = \frac{2\,N_t\,m\,v^2}{\mathcal{A}\,(v\,t)} \tag{2.45}$$

Therefore, we can relate this back to the original momentum and ultimately the temperature of the gas.

$$P = \frac{N_t\,m}{V}\left(\frac{3\,k_B\,T}{m}\right) \tag{2.46}$$

In other words, after accounting for all 3 degrees of freedom, we get

$$P = \frac{N\,k_B\,T}{V} \tag{2.47}$$

where N is the total number of gas particles in the container, k is Boltzmann's constant, T is the absolute temperature of the gas, and V is the volume of the container. Written as

$$PV = N\,k_B\,T = n\,R\,T \tag{2.48}$$

we get the familiar Ideal Gas Law where n is the number of moles of gas and R is the gas constant and we recognize that

$$N\,k_B = n\,R \tag{2.49}$$

2.1.12 Pascal Principle

As discussed in Section 2.1.11, pressure is a consequence of collisions by the fluid particles with the container wall, and the severity of that collision and subsequent change in momentum is related, in part, to the velocity of the fluid particles. Furthermore, our analysis leading to a statistical estimate of the *rms* velocity given in Equation 2.38 assumed the translational velocity in all three cardinal directions was the same due to equal probabilities. Consequently, we can conclude from these individual statements that the resulting *pressure* in each of the cardinal directions will also be the same, and this is the basis for the physical law put forward by Blaise Pascal (1623–1662) in 1653 which is now known as *Pascal's Law* or the *Pascal Principle*, which primarily relates to pressures inside a confined fluid but can also be extrapolated to open fluids such as lakes, oceans, and the atmosphere.

For a fluid at rest, the pressure at a given depth must be sufficient to support all of the weight of all of the fluid above it. Since weight is defined by Newton's Second Law given in Equation 2.1 where a is specifically the acceleration due to gravity g.

$$W = m\,g \qquad (2.50)$$

The mass of a cylinder of fluid is given by the density of the fluid, ρ by

$$m = \pi\,r^2\,h\,\rho \qquad (2.51)$$

Therefore, the weight of a column of fluid with density ρ, radius r, and height h is

$$W = \pi\,r^2\,h\,\rho\,g \qquad (2.52)$$

If we look at a submerged cylindrical sample of fluid with radius r and cylinder height x located such that the upper surface of the cylinder is at depth h below the surface of the liquid, as shown in Figure 2.6, we get the pressure at the top of the sample to be

$$P_{top} = \frac{F_{top}}{\mathcal{A}} = \frac{\pi\,r^2\,h\,\rho\,g}{\pi\,r^2} = \rho\,g\,h \qquad (2.53)$$

whereas, the pressure found at the bottom of the sample, an additional depth of x below the surface, is

$$P_{bottom} = \frac{F_{bottom}}{\mathcal{A}} = \frac{\pi\,r^2\,(h+x)\,\rho\,g\,h}{\pi\,r^2} = \rho\,g\,(h+x) \qquad (2.54)$$

so the pressure difference between the top and bottom surfaces of the fluid sample would be

$$P_{bottom} - P_{top} = \rho\,g\,(h+x) - \rho\,g\,h = \rho\,g\,x \qquad (2.55)$$

Therefore the net force acting on the sample would be

$$F_{net} = \Delta P\,\mathcal{A} = \rho\,g\,x\,\mathcal{A} \qquad (2.56)$$

which is to say

$$F_{net} = m\,g \qquad (2.57)$$

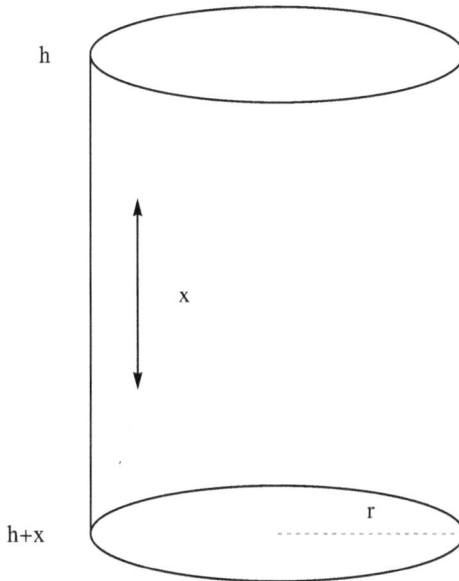

FIGURE 2.6
A cylinder of fluid with radius r and density ρ is submerged such that
the top of the cylinder is at depth h. The cylinder has height x such that
the bottom of the cylinder is at depth $h+x$.

in other words the weight of the fluid sample, itself. This is known as
Archimedes Principle.

    If we then close the top of the sample, removing all open surfaces,
the pressures at the top and bottom and all sides of the sample become
one and the same value. In other words, the pressure inside that enclosed
volume becomes uniform throughout. This is *Pascal's Principle* which
states quite bluntly, the pressure inside any enclosed fluid is the same
everywhere. This principle serves as the basis for modern hydraulics.
Since the pressure inside the hydraulic fluid is the same everywhere,
applied force becomes a function of area. The principle can be stated as
an equality of pressures applied to two different piston sizes, as shown
in Figure 2.7.

$$\frac{F_A}{\mathcal{A}_A} = \frac{F_B}{\mathcal{A}_B} \tag{2.58}$$

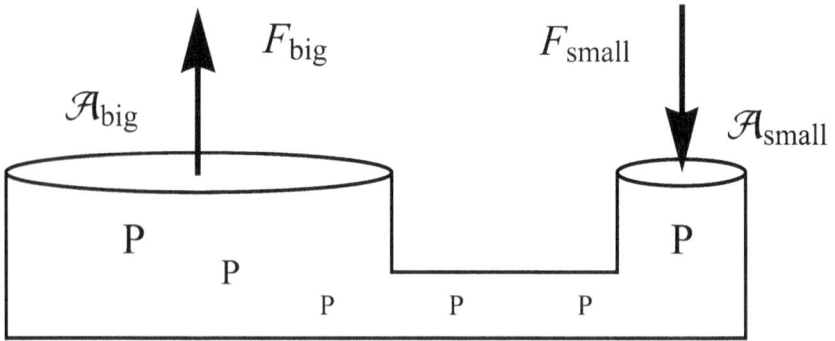

FIGURE 2.7

If an incompressible fluid is confined between two movable pistons, the amount of force will be modified based on the ratio of the surface areas. For example, a small force F_{small} acting on a small piston with area $\mathcal{A}_{small}$ will generate a larger force F_{big} when acting on a larger piston with area $\mathcal{A}_{big}$.

where F_A is the force on a piston with area $\mathcal{A}_A$ and F_B is the force on a piston with area $\mathcal{A}_B$. If the areas of A and B are not equal, as shown in Figure 2.7, hydraulic amplification can take place where the force on F_B, for example, would be

$$F_B = \frac{\mathcal{A}_B}{\mathcal{A}_A} F_A \qquad (2.59)$$

The force F_B will be greater than F_A if $\mathcal{A}_A < \mathcal{A}_B$. The bigger the surface area, the more force can be applied. This is why hydraulic pumps, which generate the pressure, have such small pistons and hydraulic rams, which do the heavy lifting, have such large pistons.

If we look carefully at the units in the equation for P_{top}, we find pressure, which is normally written as *Pascals* according to

$$[P_{top}] = [\rho]\,[g]\,[h] = \frac{kg}{m^3}\,\frac{m}{s^2}\,m = \frac{kg\,m^2}{s^2\,m^3} = \frac{N}{m^2} = Pa \qquad (2.60)$$

can be written in terms of energy as follows

$$[P_{top}] = \frac{kg\,m^2}{s^2\,m^3} = \frac{N\,m}{m^3} = \frac{J}{m^3} \qquad (2.61)$$

In other words, pressure is actually a form of energy density, in this case specifically a *potential* energy density, since it has its origins in the gravitational potential energy field.

2.1.13 Continuity Principle

Mass is one of the most fundamentally conserved quantities. In science, conservation means that all of the quantity has to be accounted for at all times and under all conditions. Such quantities are not allowed to disappear or reappear. The mass flow rate for any given fluid is given by

$$Q = \rho \left(\vec{v} \cdot \vec{A} \right) \tag{2.62}$$

where ρ is the density of the fluid, v is the linear speed of the flow, and A is the cross-sectional area through which the fluid is flowing. Continuity requires mass to be conserved through any given volume; therefore, any change in the cross-sectional area must be offset by either an increase in density, ρ, the speed v at which the fluid is moving, or some combination of the two, as illustrated by the following:

$$\left(-\frac{\Delta A}{A} \right) = \frac{\Delta \rho}{\rho} + \frac{\Delta v}{v} \tag{2.63}$$

which suggests any *decrease* in cross-sectional area must be offset by a corresponding *increase* in the density ρ, the speed v, or a mixture of both, and vice-versa. This is the basic idea behind the *Continuity Principle*. For example, if we assume for the moment that the change in density of the air remains constant while flowing through an area approximately 50% reduced in size, the flow speed would have to double.

If we look at the *divergence* of the mass flow rate Q through some cross-sectional area A, we get

$$\nabla \cdot Q = \nabla \cdot \left(\rho \, \vec{v} \, A_{\parallel} \right) = \nabla \cdot \left(\rho \, \vec{v} \right) A_{\parallel} \tag{2.64}$$

where the divergence is a spatial operator given by

$$\nabla = \frac{\partial}{\partial x} \hat{x} + \frac{\partial}{\partial y} \hat{y} + \frac{\partial}{\partial z} \hat{z} \tag{2.65}$$

Since all of the mass must be accounted for at all times, the divergence of the mass flow rate must either be 0 or accounted for by some time-rate-of-change in the density ρ at the given location indicating either an accumulation or loss of mass. In other words:

$$\nabla \cdot Q = -\frac{\partial \rho}{\partial t} \mathcal{A} \qquad (2.66)$$

Therefore,

$$(\nabla \cdot \rho)(v\,\mathcal{A}) + (\nabla \cdot v)(\rho\,\mathcal{A}) + (\nabla \cdot \mathcal{A})(\rho\,v) = -\frac{\partial \rho}{\partial t} \mathcal{A} \qquad (2.67)$$

or

$$(\nabla \cdot \rho)(v) + (\nabla \cdot v)(\rho) + = -\frac{\partial \rho}{\partial t} \qquad (2.68)$$

The term $(\nabla \cdot \rho)$ represents a change in density with respect to position, such as the atmosphere getting thinner with height or less dense in a low pressure region. In the event all other terms are 0, we get that

$$(\nabla \cdot \rho)(v) = -\frac{\partial \rho}{\partial t} \qquad (2.69)$$

which represents mass transport in or out of the sample volume. This is consistent with our discussion earlier about pressure and suggests that particles will leave high population areas for lower ones as demonstrated in Figure 2.8. In our earlier discussion, the velocity was the *rms* velocity of the gas and demonstrates how and why gas particles will expand to fill an entire volume. The gas expands until the concentration is uniform within the volume. Velocities are also distributed among the gas particles in order to evenly and uniformly distribute the energy.

The term $(\nabla \cdot v)$ would represent some change in the fluid velocity such as air leaving through a nozzle or entering into a Venturi chamber. In the event all other terms are 0, we get that

$$(\nabla \cdot v)(\rho) = -\frac{\partial \rho}{\partial t} \qquad (2.70)$$

which would indicate an increase or decrease in fluid density as a result of the spreading or constriction of the flow, as illustrated in Figure 2.9. This makes sense from the standpoint that we need to get some

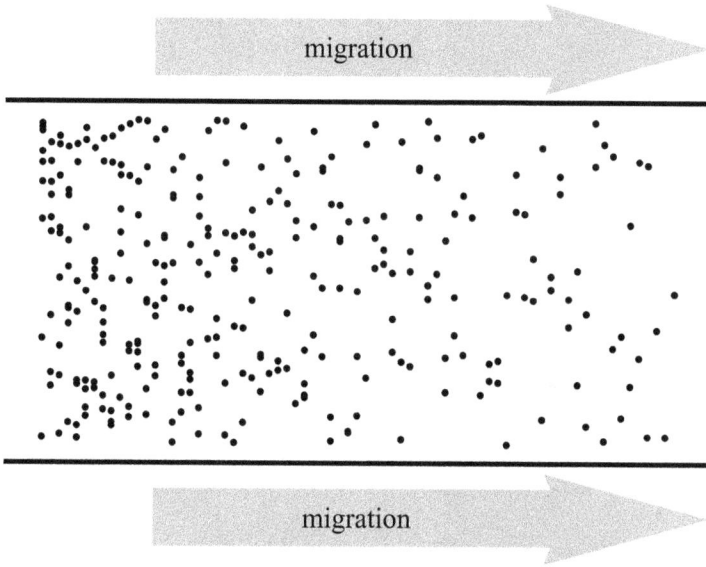

FIGURE 2.8

Because of the equal distribution of energy requirements in statistical dynamics, particles will naturally migrate from regions of high concentration to regions with lower concentrations. In the above picture, there are considerably more particles in the left half of the chamber. These particles will migrate to the right, as indicated by the arrows, until the concentration of particles is uniform throughout the chamber. This is why gas will expand to occupy the entire available volume.

number of kilograms of mass through this area. If we constrict our flow causing a velocity convergence, or negative divergence, than the density must necessarily increase to compensate. Conversely, spreading of the velocity results in a decrease in density, which is seen in the regions of low pressure in the atmosphere. The pressure drops in part due to the decrease in the number of atmospheric particles in the area. Should our fluid be incompressible, no change of density would be allowed and we get

$$\nabla \cdot v = 0 \tag{2.71}$$

which is a restatement of Pascal's Principle for a confined and incompressible fluid.

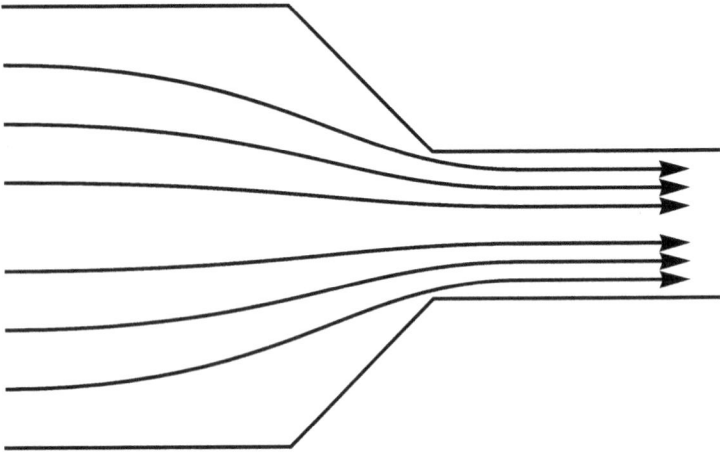

FIGURE 2.9
When a flow encounters a change in boundary conditions, a compressible fluid can increase its density or its velocity in order to maintain a constant amount of mass moving through the pipe every second. This image demonstrates the flow streams of a fluid encountering a pipe diameter reduction. Notice how the streams become more concentrated as they enter the reduced area.

If we look at the units of continuity, we get

$$[\nabla \cdot (\rho \, v)] = \frac{kg \, m}{m^4 \, s} = \frac{N \, s}{m^4} = \frac{Pa}{m \, v} = \frac{Pa}{p} \qquad (2.72)$$

or pressure per unit momentum. While this is not particularly insightful, it does demonstrate the reasoning behind continuity beyond simple mass conservation. Recall from the earlier discussion in Section 2.1.11 that pressure is a form of energy, specifically energy density. Momentum is the consequence of inertia, or mass. Energy must also be conserved which means the total amount of systemic energy at a given point in time must be constant, and that includes both *potential* and *kinetic* energies. Potential energy would refer to the density term in Equation 2.68; whereas, the kinetic energy would refer to the velocity term. In other words, to compress or expand a gas requires energy, just as changing the velocity of any fluid, compressible or not.

2.1.14 Bernoulli and Energy Conservation

Daniel Bernoulli (1700–1782) examined the total amount of energy in the flow of a fluid, including both internal quantities within the flow and external energy sources. From classical mechanics we know that the total amount of energy within a mechanical system at any given time is

$$U_{total} = W + m\,g\,h + \frac{1}{2}k\,x^2 + \frac{1}{2}m\,v^2 + \frac{1}{2}I\,\omega^2 \qquad (2.73)$$

if we divide each term by a sample volume $\mathcal{V}$ we get

$$\frac{U_{total}}{\mathcal{V}} = \frac{W}{\mathcal{V}} + \rho\,g\,h + \frac{k}{2\,\mathcal{V}}x^2 + \frac{1}{2}\rho\,v^2 + \frac{I}{2\,\mathcal{V}}\omega^2 \qquad (2.74)$$

We have already shown that work per unit volume is a form of pressure. The second term we recognize from Pascal's Principle. The spring term would apply to viscoelastic fluids. The velocity term is the energy in the linear flow; whereas, the ω term would be the rotational energy in the fluid. This rotation is not necessarily due to turbulence. It could simply be the nice, uniform rotation of a fluid like stirring a cup of tea.

If we confine our discussion to normal fluids, we are left with

$$\frac{U_{total}}{\mathcal{V}} = P + \rho\,g\,h + \frac{1}{2}\rho\,v^2 + \frac{I}{2\,\mathcal{V}}\omega^2 \qquad (2.75)$$

Even though Bernoulli did not expressly discuss viscous fluids, this does give some insight into the issues surrounding turbulence. As a fluid becomes turbulent, it starts developing small vortices near the boundary walls within the boundary layers. Each of these vortices contain a certain amount of rotational energy as given by the ω term. As suggested by the moment of inertia, I, the amount is going to depend on both the speed of rotation and the size of the vortex. Unfortunately, Bernoulli's description does not give us a mechanism to determine the correlation between the linear or laminar velocity v and the rotational velocity ω. In fact, most would presume the ω in this formula would strictly be the collective rotation of the entire fluid, but that is a very narrow interpretation.

If we disregard the rotational issues and confine ourselves to laminar or irrotational flow for now, we are left with

$$\frac{U_{total}}{\mathcal{V}} = P + \rho\,g\,h + \frac{1}{2}\rho\,v^2 \qquad (2.76)$$

According to Bernoulli and the principles of energy conservation, this quantity must be a constant. Again, it is important to remember that Bernoulli did not consider viscous forces; therefore, some energy may be lost to the fluid itself in the case of *real* fluids. Much of this is likely to go to thermal energy in the form of an increasing temperature, but a significant portion could also go to changing other characteristics of the fluid medium, such as the viscosity or density.

Several important relationships come from a careful examination of Equation 2.76. If we further consider an incompressible fluid over a short distance, so the change in elevation is negligible, we see that the change in velocity is offset by an opposing change in pressure, as shown by

$$\frac{1}{2}\rho \, \Delta\left(v^2\right) = -\Delta P \qquad (2.77)$$

or written another way

$$P_f - P_i = \frac{1}{2}\rho \left(v_i^2 - v_f^2\right) \qquad (2.78)$$

where P is the local fluid pressure, ρ is the density of the fluid, and v is the velocity or speed of the fluid stream at the given location. Notice the order of subtraction is reversed, therefore an increasing velocity, where $v_f > v_i$, results in a negative pressure change such that $P_f < P_i$. This is known as the *Venturi Effect*. In other words, increasing the flow velocity corresponds to a decrease in pressure. It is interesting to note, Bernoulli's explanation of the Venturi Effect works to high precision, even in the classical demonstration with a high velocity turbulent flow of air over a piece of paper, *in spite of the fact* that neither air nor water are *inviscid* fluids. This demonstrates that Bernoulli's Equation is useful for an initial examination, at least, even though it was not intended for viscous let alone turbulent flows.

2.2   Viscous Fluid Flow

Viscosity is an inherent property of any real fluid. Unfortunately, unlike mass or temperature, viscosity is not a quantity that is understood from first principles and cannot be analytically predicted with any precision whatsoever.[2] Consequently, the viscosity of a fluid must be empirically measured for each given fluid or mixture. Unfortunately, even this process has its challenges, since not all fluids respond to a driven flow in the same way. There are currently more than a dozen recognized curve-fitting models in common use today with the number of fitting parameters ranging from 1 to 9. Some of the models include Newtonian, Cross, Carreau-Yasuda, Power Law, Herschel-Bulkley, Bingham Plastic, Arrhenius, Intrinsic, Huggins, and Mark-Houwink-Sakurada. In addition to the initial curve fitting parameters, Some of the models have baseline parameters within them which also require several curve fitting parameters. The Cross model is one such example, where the viscosity curve depends on a term call η_0 which in turn requires a two-parameter curve fit.

2.2.1 Viscosity: Definition and Measurement

All viscosity measurements are curve fits between the quantities of *shear stress* and *strain rate.* Some may also include temperature or pressure, but we will not be considering those types of fluids. In this discussion we are going to consider Newtonian fluids that are direct correlations between shear stress and strain rate.

A fluid sample is placed between a rigid boundary and a moving surface, as shown in Figure 2.10, and a controlled amount of force is used to constantly shear the fluid between the two surfaces. Strain is the amount of deformation a substance would have to a given strain force. Unlike solids, fluids cannot sustain a fixed amount of shear; therefore, the strain rate is the amount of deformation per second of the fluid, in other words, the flow rate.

Isaac Newton first postulated a linear response by a fluid to a driven shear force, and a large number of fluids fit that description quite well. As a result, these fluids are often referred to as *Newtonian Fluids,* since their viscosity can be described by a simple, linear fit. That means the

---

[2]There have been three such attempts in history, all involving liquid Argon. The initial attempt was off by three orders of magnitude. The last and closest attempt overestimated the viscosity by more than one order of magnitude.

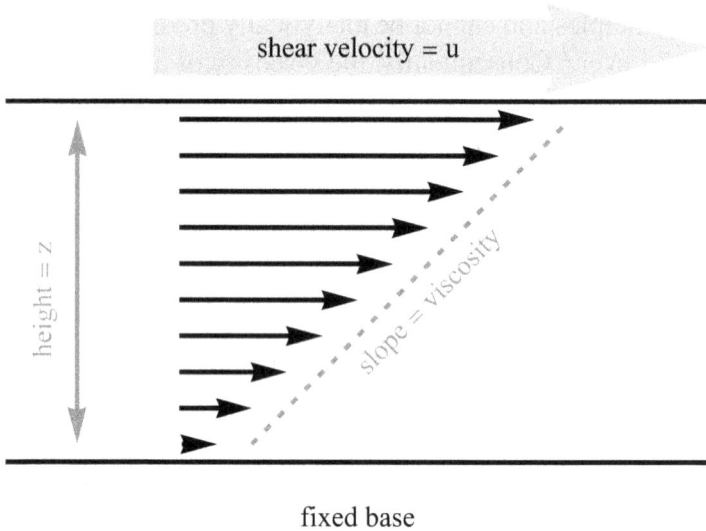

FIGURE 2.10
For a Newtonian Fluid, a sample is placed between a rigid base and a movable plate. The gap distance is z, and a shear force is placed on the upper surface. The amount of strain that results per second on the upper surface is measured along with the amount of stress required to keep a constant velocity. The slope of the line that results from a series of measurements at different stresses and strain rates results in an empirical measurement of the viscosity of the given fluid.

strain rate is *directly proportional to* the shear stress.

$$\sigma = \mu \frac{d\varepsilon}{dt} \tag{2.79}$$

where σ is the shear stress, $\frac{d\varepsilon}{dt}$ is the rate of deformation or strain rate, and μ is the viscosity of the fluid. In terms of measured quantities, this can be written as

$$\frac{F}{A} = \mu \frac{u}{z} \tag{2.80}$$

where F is the applied and measured force driving the fluid, A is the surface are a over which the force is applied, u is the speed at which the top layer of fluid is moving, and z is the thickness of the fluid layer.

Water is the standard for viscosity and has a value of precisely $0.01 \frac{g}{cm\,s}$ or 1 *centipoise* (*cP*), named after Jean Poiseuille (1797–1889). Other standard units include the Pascal-Second (*Pa · s*) which is given by

$$1\,Pa\,s = 1\frac{N\,s}{m^2} = 1\frac{kg}{m\,s} = 10\frac{g}{cm\,s} = 10\,Poise = 1000\,cP \quad (2.81)$$

Therefore,

$$1\,cP = 0.01\,Poise = 1\,mPa\,s \quad (2.82)$$

By comparison, air has a viscosity of approximately 1.73×10^{-3} *cP* or perhaps 1.73 *μPa s*, and the glycerin that is used in face cream has a viscosity of roughly 1000 *cP*.

The viscosity concept itself refers to the ability of the fluid to adhere to itself, which is why "thicker" fluids like peanut butter and molasses seem more sticky and pour more slowly than lower viscosity fluids such as rubbing alcohol and water. This cohesion is different than the bonding of atoms, as demonstrated by mercury which has an extremely high internal bond strength as indicated by its huge surface tension but nevertheless has a viscosity comparable to water at about 1.55 *cP*. Unlike friction, viscosity is more about deformation than destruction: how hard is it to cause the material to deform. In the case of air, the following formula has been worked out empirically for the viscous drag of air flowing through a pipe:

$$\Delta P = \left(7.57 \times 10^4\right)\frac{L\,Q^{1.85}}{d^5\,P} \quad (2.83)$$

where the pressure drop ΔP and initial driving pressure P are in kg/cm^2, the pipe length L is in *meters*, the volumetric flow rate Q is in m^3/min, and the pipe diameter d is in *mm*. The strange exponents and odd mixture of units in this empirical model indicates just how experimentally driven much of fluid dynamics has become due to an understandable but unfortunate theoretical vacuum in the wake of impossibly complex and highly coupled four-dimensional differential equations.

2.2.2 Poiseuille's Law and Laminar Flow

Flow through any arbitrary space can be viewed as the movement of adjacent layers in the direction of the flow velocity, like so many

3x5 cards in a stack. The middle cards move adjacent to each other, and the viscosity causes each layer to "pull" the adjacent layers along with them. If the flow is bounded by a rigid surface, such as the wall of a pipe or rectangular channel, the velocity will decrease as the layers get closer to the rigid surface, as illustrated in Figure 2.11. The **no-slip** condition requires that the flow velocity immediately adjacent to such a rigid boundary must identically match the velocity of the rigid surface. In the case of a pipe or channel, that typically means 0, so the flow will be maximum in the center and decrease gradually until it achieves 0 at the wall of the pipe.

Consider a typical viscous flow through a pipe with radius R, such as shown in Figure 2.11. Fluid dynamics is governed by the Navier-Stokes

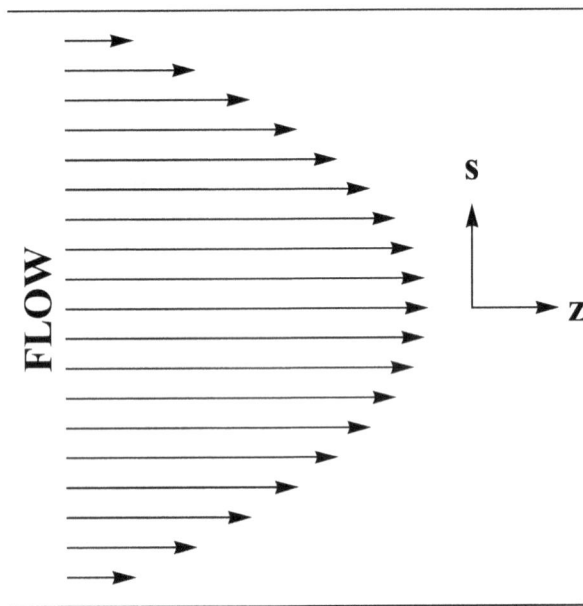

FIGURE 2.11

Laminar Pipe Flow is demonstrated by adjacent layers all moving in the same direction. Flow is maximum in the center and decreases across the radius of the pipe until it reaches 0 at the wall.

Equation,

$$\frac{\partial}{\partial t}(\rho v) + \vec{v} \cdot \left(\vec{\nabla} v\right) = \Delta\rho \, \vec{g} - \vec{\nabla} P + \eta\nabla^2 v \qquad (2.84)$$

where ρ is the density of the fluid, $\vec{v}$ is the flow velocity, $\vec{g}$ is the acceleration vector of gravity, P is the pressure on or in the fluid, and η is the fluid viscosity.

The first term describes the change in flow or fluid density over time. If we confine ourselves to fully developed *steady states,* where the flow does not significantly change with time, this first term will vanish. If we further assume, for the sake of this consideration, a uniform density and incompressible flow, we can further neglect the gravitational influence on the right hand side.

The "non-linear" term, $\vec{v} \cdot \left(\vec{\nabla} v\right)$ represents inner circulation within the pipe due to a turbulent boundary layer. If we constrain ourselves only to laminar flow, then this term is approximately 0; however, if the flow becomes turbulent then this term becomes significant as eddy currents and swirling takes place, and the flow becomes highly variable. We will discuss turbulence and the development and growth of the boundary layer in the next section. For now, we will assume laminar flow. That leaves the following terms:

$$\eta\nabla^2 v = \vec{\nabla} P \qquad (2.85)$$

which gives

$$\frac{\partial P}{\partial s}\hat{s} + \frac{\partial P}{\partial z}\hat{z} = \eta \left[\frac{1}{s}\frac{\partial}{\partial s}\left(s\frac{\partial v}{\partial s}\right) + \frac{\partial^2 v}{\partial z^2}\right] \qquad (2.86)$$

If the flow is laminar, there is little or no radial flow, so we can assume $v_s \approx 0$. We can then examine the $\hat{z}$ flow, which is the flow down the pipe. The differential equation we must consider is the following:

$$\frac{\partial P}{\partial z} = \eta \left[\frac{1}{s}\frac{\partial}{\partial s}\left(s\frac{\partial v_z}{\partial s}\right) + \frac{\partial^2 v_z}{\partial z^2}\right] \qquad (2.87)$$

If we further assume a constant pressure gradient in the $\hat{z}$ direction

$$\Delta P = \frac{\partial P}{\partial z}$$

along a pipe of length L, then Equation 2.87 has the following solution:

$$v_z(s) = \frac{\Delta P(s^2 - R^2)}{4L\eta} \qquad (2.88)$$

which, as expected, is maximal at the center of the pipe and 0 at the walls. We can average this flow by integrating over the cross section of the pipe to find the average volumetric flow rate as

$$Q = \frac{\Delta P \pi R^4}{8L\eta} \qquad (2.89)$$

In this derivation we have only considered laminar flow. If this is truly the case, Poiseuille's Law as shown in Equation 2.89 should hold fairly well; however, if the flow is turbulent and Poiseuille's Law as written in Equation 2.89 is no longer valid.

2.2.3 Development of the Boundary Layer

If a flow is driven too hard, the shear between adjacent fluid layers becomes too great and buckling will occur. When this happens, the nonlinear term in Equation 2.84 becomes significant and the adjacent fluid layers no longer move uniformly in the $\hat{z}$ direction down the pipe and begin to curl toward the stationary walls of the pipe or channel. This condition is known as *turbulence,* and it can greatly impede the flow. Indeed, it puts an upper limit on the volumetric flow rate Q since any additional energy expenditure in the form of pressure only goes to reinforcing the turbulence in the boundary layer and generates little or no change in Q.

The Reynolds Number is used to estimate the type of flow that would occur under given circumstances and is given by the following expression:

$$Re = \frac{\rho v L}{\mu} \qquad (2.90)$$

where ρ is the density of the fluid, v is the linear speed of the flow, and L is typically the diameter of the pipe or channel. As outlined in Section 2.2.2, laminar flow is characterized by adjacent layers moving in parallel with each other. However, boundaries such as the sides of a pipe or flow channel create variations in the speed of the flow. If the driving pressures remain modest, the flow layers remain parallel as they move down the pipe and around obstacles like cars and airplane wings. However, if the flow speed increases too much, these adjacent layers can start to pull on one another and the flow will transition to a state of turbulence, characterized by swirling regions like miniature **hurricanes** called *vortices,* and boundary layers of trapped fluid can develop along the interior walls of the pipe. The Reynolds Number gives us a way to estimate the maximum stable velocity v and resulting volumetric flow rate Q.

The Reynolds Number is a dimensionless quantity relating the inertial (mass) and viscous forces acting within a given fluid. Unlike other physical laws and quantities, the actual value of the Reynolds Number is of less significance than its order of magnitude. Adding to the confusion, transitions from laminar to turbulence depend greatly on the conditions of the boundary walls. If the wall is considered "smooth" and free of defects that might disrupt the layers, the flow might remain laminar for Reynolds Numbers as high as 2000 or more; however, if the walls of the pipe are rusted and rough, the flow might transition to turbulence with Reynolds Numbers as low as 50. Laminar flow can only be guaranteed if the Reynolds Number is less than 1, and turbulence is virtually guaranteed for Reynolds Numbers greater than 5000.

Once the transition to turbulence is triggered, the boundary layer will grow larger as the flow moves down the pipe in the z direction, but it does not grow indefinitely. Rather, the growth follows an inverse power law in z similar to

$$\frac{\delta}{z} = \frac{c}{Re_z^{1/5}} \tag{2.91}$$

where δ is the boundary thickness, c is some constant, and Re_z is the Reynolds Number with z as the length unit

$$Re_z = \frac{\rho \, v \, z}{\eta} \tag{2.92}$$

Putting these two together results in

$$\delta = c \left(\frac{z^4 \eta}{\rho v} \right)^{1/5} \tag{2.93}$$

Opinions vary considerably on the magnitude of the power law, but most documentation use either $1/5$ or $1/7$ powers for turbulent pipe flow. Furthermore, c can take on different values, as well, such as 0.38 or 0.55. Different fluids and different surfaces will obviously respond differently under different conditions, and with no clear theoretical explanation for viscosity, it is hardly a surprise that different models and parameter values are required for different situations.

The details of boundary layer growth and its dependency on the velocity is secondary to the conceptual understanding of the consequences of this boundary layer. Not only does it impede the pipe flow by occupying some amount of area, and consume more and more of the applied energy trying to drive the flow through the pipe, but it also changes the inner flow's boundary conditions by providing a buffer region between the rigid boundary surface and a fluid-type boundary that is able to match the flow of the fluid. Consequently, whatever flow energy the inner fluid might possess is uniformly distributed across the central flow creating a uniform flow velocity, as illustrated in Figure 2.12. While a uniform flow velocity might seem beneficial, one must not forget the occupied areas covering the surrounding walls have reduced the overall net cross-sectional area available for this uniform flow, thus the net volumetric flow rate is much less that would be possible were the entire pipe area available, and still less than the laminar flow equivalent. Turbulence is bad, and there is no getting around it. Unfortunately, it is also often unavoidable, which is why it still garners such attention.

2.3   Compressible Flow

Complex fluid dynamics is often difficult, if not outright impossible, to resolve analytically. Compressible fluid dynamics is almost always impossible to resolve analytically; therefore, computational methods far beyond the scope of this text are frequently employed to resolve these challenges. While these methods are able to provide a solution for a given configuration, they lack the generality that an analytical solution might provide, as well as situational insight regarding the influence or

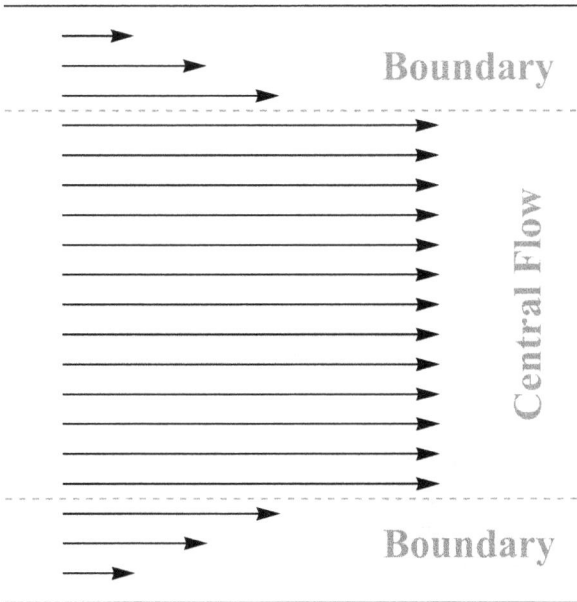

FIGURE 2.12
Turbulent Pipe Flow is demonstrated by turbulent layers that consume large amounts of flow energy and confine the laminar flow region to a smaller cross-sectional area of the pipe. Flow is uniform in the central region and decreases rapidly in the boundary layers reaching 0 at the wall.

significance of individual parameters or dimensions. Consequently, one often finds themselves re-running computational solutions numerous times and adjusting the different input parameters in order to gauge the results. Considering some solutions may take hours or even days to resolve to an acceptable degree of accuracy, this can be an extremely time consuming process.

The physics of respiration are a direct application of compressible-complex fluid dynamics. This is an extremely difficult field of soft-matter physics due to the compressibility of the fluid, namely air, which may cause the density of the fluid to vary from one location to another, and the fact the fluid is complex, meaning it is not a pure and single species of one particular molecule. In fact, as will be discussed later in Section 3.1.1, the body goes out of its way to make the air in the respiratory system complex by adding moisture, i.e., water, so in addition to having oxygen, nitrogen, argon, neon, krypton, and all the other species of gas molecules, we also have liquid water suspended in this mixture of gases.

In the case of a compressible fluid, such as air, a certain amount of potential energy can be expended or withdrawn in generating regions of high or low pressure, respectively. It is the very fact that air is compressible that makes level flight possible, since a region of high pressure underneath the wings can be created that perfectly offsets the weight of the aircraft. In an incompressible fluid, such stable pressure regions are not possible and generating a stable lift force and level trajectory is much more complex.

Variations in density arise in two locations in Equation 2.84. The first is the time-dependent term which looks at the density change over time at a specific location. This would represent an accumulation or removal of fluid material from this location. For example, if we take the flight of an irrotational baseball, the air accumulates immediately in front of the ball as it moves through the viscous fluid. The air is also stretched and decompressed behind the baseball, since the ball previously occupied that physical space. When the ball moved, the surrounding air is compelled to move inward and fill the void left by the baseball. That takes time and, in the meanwhile, there is a lack of fluid material in this area. These two factors generate a huge pressure gradient across the diameter of the baseball resulting in a huge *drag force* that causes the ball to slow down rapidly. This is why the *knuckleball* is the slowest pitch in baseball.

If we only take the density and pressure terms from Equation 2.84 and rearrange them slightly, we get

$$\vec{\nabla} P = \Delta\rho \, \vec{g} - \frac{\partial}{\partial t}(\rho v) \tag{2.94}$$

The first right-hand term is basically the Archimedes Principle as derived back in Equation 2.57. This deals more with a *difference* in density rather than a *change* in density. Since we have already discussed this, and we are more interested in flow-generated density changes right now, we can safely set that aside. That leaves

$$-\vec{\nabla} P = \rho \frac{\partial v}{\partial t} + v \frac{\partial \rho}{\partial t} \tag{2.95}$$

so a pressure gradient can generate either a variation in speed, which is consistent with Newton's Second Law, or a variation in density, which is consistent with Pascal's Law, or some combination of the two. In fact, in the case of the baseball, it is the pressure gradient created by the baseball's velocity and the accumulation in front and dispersal of mass behind the ball that generates the acceleration, or change in velocity, of the air as it passes around the ball. It is the penultimate statement of pressure driving and guiding both fluid flow and mass distribution.

3

Pulmonary Design and Operation

The functional purpose of the pulmonary (respiratory) system is the movement of significant quantities of air and the exchange of gases. To accomplish this goal, the system is made of four main parts: an airway to facilitate transport of the physical mass in and out of the body, a multi-stage filtration system to rid the air of damaging or harmful materials such as dust and large water droplets, a pressure system to induce movement of the compressible fluid, and the gas exchange membranes which connect the pulmonary system to the circulatory system allowing oxygen to reach the blood and waste gases to leave. Most of these main components each have multiple parts, so we will take each one of these in turn in order to gain an overall picture of the working system as a whole.

3.1 The Upper Airway

The human airway is basically a series of connected tubes with the open atmosphere on one end and the lungs on the other collectively known as the *respiratory tract*. The airway is divided into the upper and lower parts roughly split at the larynx, or voice box. The upper part of the airway includes the nose and nasal cavity along with the mouth and oral cavity. These two cavities merge into what is called the pharynx, or throat cavity, which leads into the larynx and the lower tract starting with the bronchus and subsequently the bronchioles, subdividing into smaller and smaller tubes as they enter the region known as the lungs. The smallest tubes, called terminal bronchioles, end in small air sacks called alveoli which is where the gas exchange process takes place.

DOI: 10.1201/9781003683476-3

Nose and Nasal Cavities

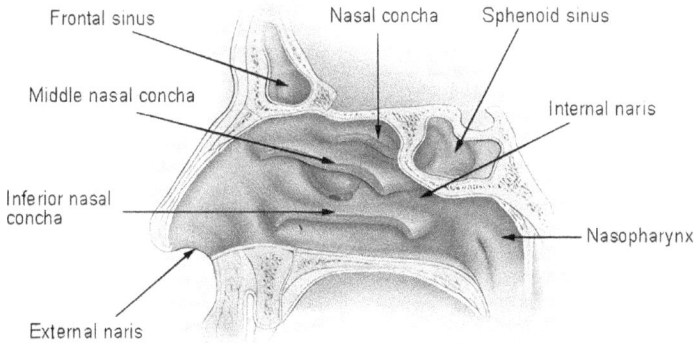

FIGURE 3.1

The nasal cavity starts with the external naris, or the nose, and extends back to the nasopharynx. The chamber is divided into different sections by the nasal concha or turbinates. The interior lining of the nasal cavity is coated with a thick mucous which both humidifies the air as it passes and traps contaminants before they can enter the main airway. (Centers for Disease Control & Prevention, part of the United States Department of Health & Human Services.)

3.1.1 Nasal Filtration

The interior lining of the left and right nasal cavities as far back as the pharynx contain a high concentration of mucous-generating glands. The tissue also contains a high concentration of blood vessels which warm any air that comes in contact with these surfaces. The mucous glands not only make sure that the incoming air is humidified, but the sticky surface also traps pollen, dust, and other particles and disease-carrying material that might be suspended in the air. The trapped material might later be expelled during the exhalation process or by "blowing your nose."

To help facilitate the filtration, warming, and humidification process, the left and right nasal cavities are further subdivided by three horizontal sections by three *conchae,* also known as *turbinates* as shown in Figure 3.1. These curved horizontal plates extend from the outer sidewalls inward toward the nasal septum and reduce the effective airway diameter which increases the airspeed through the region in order to preserve the mass flow rate as outlined in Section 2.1.13.

According to a study by the National Institute of Health (NIH) [3], human females have nasal cavity sizes from $19\,cm^3$ to $26\,cm^3$, whereas human males often have larger cavities ranging from $19\,cm^3$ up to $31\,cm^3$. Within this volume, the conchae are not uniformly spaced, with the bottom or inferior conchae being farthest from the lower wall and the upper or superior concha being closest to the upper wall. This means the lower passage, or *meatus*, is generally considered the largest of the airways through the nasal cavity. The lower passages contribute the most to respiration, while the upper passage contains the olfactory nerves which provides the sense of smell. The actual thickness of these conchae, and how much they restrict the airflow, varies with age and physical condition. Allergies and colds can cause inflammation and swelling of these tissues reducing the area available for airflow, which is why a person would feel "stuffed up" when they are suffering from seasonal allergies. However, if we assume for the sake of simplicity, the airways are of equal size obstructing half of the flow area, and the cavity is an average $25\,cm^3$ and roughly cubic in shape, our nominal and unobstructed flow passage would have approximately $8.55\,cm^2$ in cross-sectional area. The conchae would collectively reduce that to $4.27\,cm^2$.

According to the NIH National Library of Medicine, the average human circulates an average of $500\,mL$ of air with each breath[1] while resting, a quantity known as the *tidal volume.* This equates to $500\,cm^3$ of air with each breath, and the average respiration rate for an adult at rest is approximately $20/s$. According to Equation 2.62, that yields a flow speed v of approximately $39.0\,cm/s$ in the area between the nasal conchae. In the case of air moving through a cross-section of $4.27\,cm^2$ at a speed of $1.17\,m/s$, Equation 2.90 gives a Reynolds Number of

$$Re = \frac{(1\ kg/m^3)(39.0\ cm/s)(\sqrt{4.27\ cm^2})}{1.73 \times 10^{-5}\ \frac{kg}{m\,s}} \approx 466 \qquad (3.1)$$

The vortices that develop along the surfaces of the conchae and nasal cavity serve to recirculate the air, repeatedly bringing the air in contact with the warm and moist surfaces which both warms the air and increases the humidity or moisture content within the air. These vortices

---

[1] Physiology, tidal volume.

also drive the heavier contaminants against the wall due to centrifugal force. If we assume the thicknesses of the boundary layers are comparable to the radius of the passageway, and the midline or tangential velocity is comparable to the flow velocity, then Equation 2.11 gives a centrifugal force of

$$r \approx \frac{\sqrt{4.27 \, cm^2}}{2} = 1.04 \, cm \tag{3.2}$$

Assuming a central velocity of $1.17 \, m/s$ would give an angular velocity of

$$\omega = \frac{v}{r} = \frac{1.17 \, m/s}{1.03 \, cm} = 114 \, s^{-1} \tag{3.3}$$

which is more than $1000 \, rpm$. At this rotation rate, the centrifugal force driving the pollen against the wall, as given by Equation 2.11, would be more than $100 \, N/kg$. By way of comparison, Earth's gravity is approximately $10 \, N/kg$, so this centrifugal force is roughly ten times that of Earth's gravity. In this way, the nasal passageways are able to remove objects up to approximately $10 \, \mu m$ in diameter while simultaneously warming and humidifying the air in preparation for the lower and more sensitive respiratory areas.

3.1.2 Mouth Breathing and the Throat

Breathing through the mouth bypasses all of the conditioning and filtering of the nasal passages. The purpose of the mouth is more directed toward the digestive systems, as evidenced by the existence of teeth, taste sensors, and salivary glands, all of which are related to the pre-processing of foods. Nasal passages are largely designed to process incoming air, with a side-note for the sense of smell just as the mouth has sensors for taste.

Nevertheless, the oral and nasal passages merge at the back of the throat in an area called the *laryngopharynx* and head down toward the larynx, or voice box, as shown in Figure 3.2. However, these two passageways unite for only a few centimeters before spitting again into the *esophagus,* or "food tube," and the *trachea,* or "wind pipe." Because the two passages both merge and divide, either can be used for both food or breathing purposes. Indeed, "feeding tubes" are often inserted through the nasal cavity, down through the laryngopharynx, and into the

Pharynx

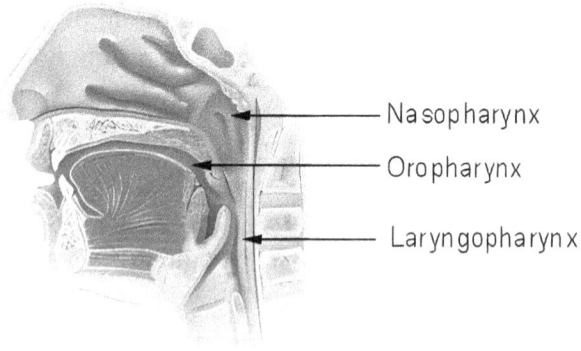

Nasopharynx

Oropharynx

Laryngopharynx

FIGURE 3.2
The pharynx is divided into three parts as shown in this figure. The upper part is the nasal pharynx which joins with the nasal cavity. The middle part is the oral pharynx which joins with the oral cavity, or mouth. The bottom part is the laryngopharynx which joins the upper two segments with the main airway leading into the larynx. (Centers for Disease Control & Prevention, part of the United States Department of Health & Human Services.)

esophagus. Alternatively, "breathing tubes" are often inserted through the much larger oral cavity, through the laryngopharynx, and into the trachea. During normal operations, the body directs the flow of either food or air using a valve known as the *epiglottis*. For breathing to occur, the epiglottis must be open allowing access to the trachea. When you swallow, the epiglottis folds down over the trachea directing the liquids and solids into the esophagus and subsequently the stomach. Choking occurs when the swallowing process is incomplete for some reason, and the food or other ingested object blocks the epiglottis from opening again. Since the epiglottis closes by folding down, rescue techniques which increase the air pressure in the windpipe assist in the upward opening motion in addition to helping drive the obstruction back into the oral cavity and out of the way.

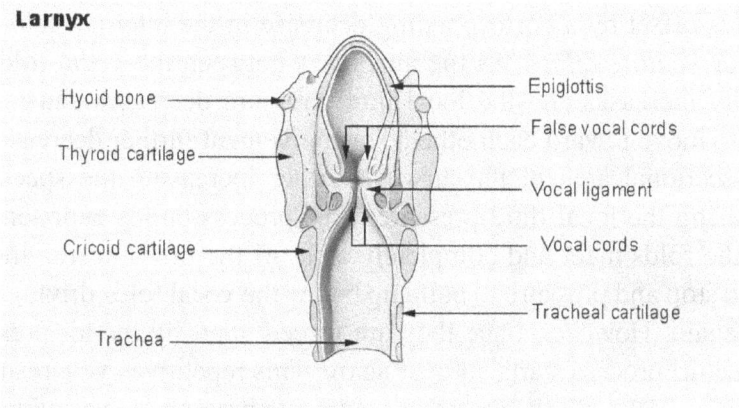

FIGURE 3.3
The larynx is a cartilage encased chamber commonly known as the voice box. The chamber truly is a box, roughly square, which contains ligaments and muscles which control the vocal cords. (Centers for Disease Control & Prevention, part of the United States Department of Health & Human Services.)

3.2 The Larynx

Commonly known as the *voice box,* the larynx is a cartilage encased chamber located at the top of the trachea just below the point where the esophagus branches off and serves as the dividing point between the upper and lower airways. The chamber truly is a type of box which contains a number of ligaments, muscles, and a pair of softer tissues called "folds," as shown in Figure 3.3. The upper fold is called the vestibular folds, and its purpose is unclear; however, the lower folds are the ones known as the "vocal cords" or vocal folds and is directly responsible for generating sound. The fold is connected by the vocal ligaments to the vocalis muscle which controls the tension in the vocal folds, thereby altering the frequency or pitch of the sound generated, if any.

Sound is generated by the vocal fold as air passes between the folds. As the air passes through the *glottis,* the narrowest region in the larynx bounded by the vocal folds, the air is forced to pick up speed according to the conservation principle outlined by Equation 2.62. Bernoulli's Equation given in Equation 2.78 shows that such an increase in flow speed is offset by a corresponding decrease in the internal fluid pressure, or *Venturi Effect.* As the air passes between the vocal folds, its velocity increases and the local fluid pressure decreases causing the folds to move toward each other. This movement further decreases the cross-sectional area of the airway, further increasing the speed and decreasing the local fluid pressure. This process builds exponentially until the folds meet and completely shut off the flow causing the airflow to stop and pressure to build up below the vocal folds driving them open again. However, once they are forced apart by the air pressure, the Venturi process starts all over again. This repetitive cycle results in vibrations in the glottis generating pulsating pressure waves within the larynx, in other words, sound waves. Adjusting the tension in the vocal folds changes the rate at which this process operates, in other words the frequency of the sound produced.

If the sound generated by the vibration of the vocal chords was the end of the story, it would be very difficult to communicate since the amplitude of the sound would be very small; however, the sound waves set up resonance patterns in the upper part of the larynx, called the *vestibule,* and subsequently in the throat and oral cavity where the sounds are further amplified and modified. The creation and amplification of the sound by the resonance within these different chambers is called *phonation,* whereas the manipulation of the sound into words is called *articulation* and is accomplished with the skillful use of the tongue, teeth, and lips.

Sound waves will resonate in a chamber provided the wavelength is an integer multiple of a characteristic length of the chamber, such as the diameter or height, as shown in Figure 3.4 and given by

$$n = \frac{2L}{\lambda}; \quad n = 1,2,3,4,\dots \tag{3.4}$$

where L is the length of the physical dimension, λ is the wavelength of the sound, and n is any positive integer starting with 1. What we know as *pitch* is actually the *frequency* of the sound wave, f.

As discussed in Section 2.1.8, the sound speed depends only on the temperature and humidity of the air, both of which should be approximately constant inside the body; therefore, the frequency and wavelength will be inversely related. Furthermore, the wavelengths will be governed by those frequencies that will *resonate* within the vocal chambers, and that means they are restricted by the physical

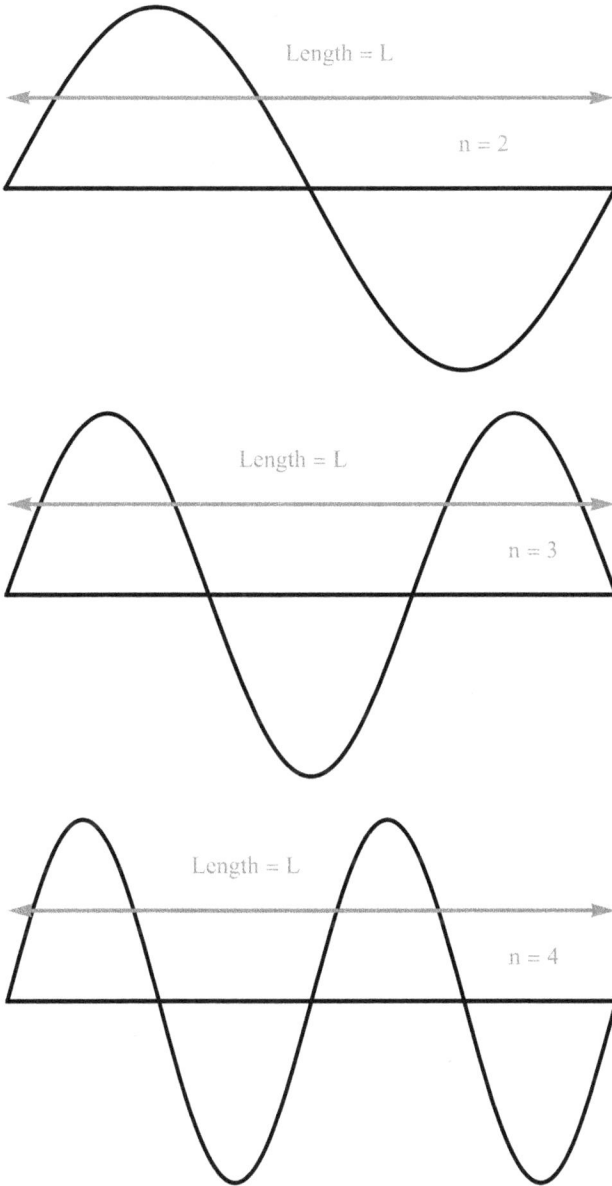

FIGURE 3.4
Sound will resonate in a cavity provided the wavelength of the sound is an even integer multiple of the wavelength. The image shows three such resonant frequencies in a chamber of length L. These plots represent the 2nd, 3rd, and 4th harmonics and have wavelengths equal to L, $2\,L/3$, and $L/2$.

dimensions, particularly that of the larynx itself. Newborn infants are relatively small, consequently their larynx is also very small, and their voices tend to have a relatively high pitch. As the child grows, their physical dimensions increase, including the size of the larynx, and their voices change over the years, generally decreasing in pitch as the wavelength selections increase in size. Adult males typically have a larger airway and larger larynx size and therefore lower voices than adult females. A recent study by the Zhang[7], which was based in part off observations by Titze[5] and others, indicates the vocal tones and resonances track directly with vocal fold lengths and other physical parameters, with females demonstrating fold lengths nearly half that of males. This is the primary reason females generally have a higher pitched voice than males.

3.3   The Lower Airway

The trachea is a large reinforced tube extending from the larynx in the neck to about the middle of the chest cavity, thereby constituting the majority of the lower airway. It is roughly an inch ($25\ mm$) in diameter and encircled with rings of rigid cartilage which gives it some structural integrity to withstand both positive and negative pressures inside and out, a characteristic that will be of significance later when we discuss the respiratory process itself. The interior of the trachea is lined with small hairlike structures and mucous glands known as the *respiratory epithelium.* As the air swirls its way up and down the trachea, particulates in the air get stuck to these mucous-coated hairs. The entire structure in constantly moving in an upward trending wave pattern which constantly "sweeps" the air clean of contaminants and then migrates those items upward toward the larynx for eventual expulsion, probably through exhalation or a stimulated cough reflex. These filaments are much finer and stickier than the nasal membranes and capable of trapping and removing particles down to 1 μm in diameter.

At the bottom of the trachea, the airway splits into a left and right branch. It is interesting to note, for most people the left branch is slightly smaller than the right, and the right branch has less angular deviation whereas the left branch is a sharper turn. A study by Daroszewski et al.[2] found the right bronchial angle averaged 26.9° whereas the left bronchial angle averaged 64.8°, more than twice as sharp. Consequently, the angle between the left and right branches averaged 73.1°, less than a score of degrees short of being a right angle. Furthermore, the average diameter of the left bronchus measures approximately 13 *mm*, whereas the right bronchus has an average diameter of 16 *mm*.

The right bronchus runs only about an inch before branching, whereas the left runs nearly twice as far. Nevertheless, both the left and right primary branches then rapidly start branching over and over getting progressively smaller and smaller each time forming what is called the *bronchial tree* due to its shrub-like appearance as shown in Figure 3.5. The lungs themselves are divided into segments known

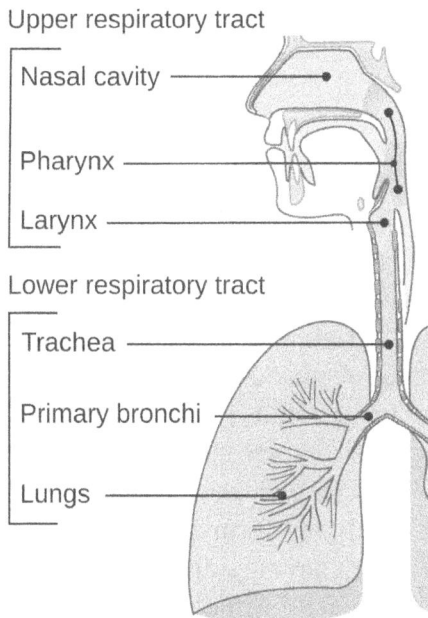

FIGURE 3.5

The lower airway starts with the larynx. The trachea connects the larynx to the primary bronchi where the airway starts repeatedly branching into smaller and smaller airways as they move into the lungs. (Centers for Disease Control & Prevention, part of the United States Department of Health & Human Services.)

Bronchi, Bronchial Tree, and Lungs

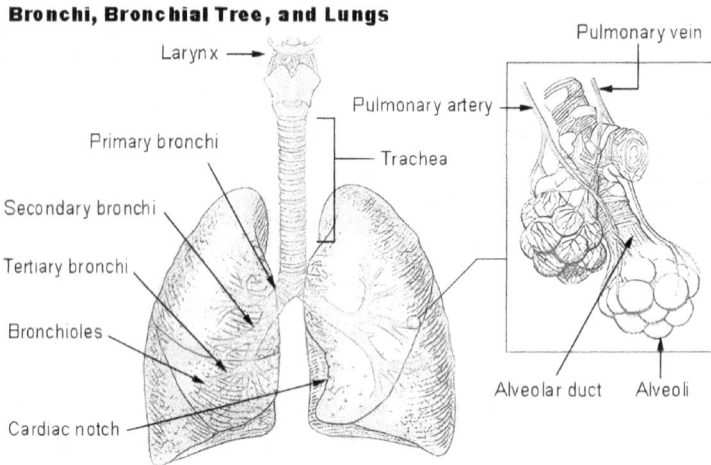

FIGURE 3.6
The primary bronchi branch into secondary and tertiary bronchi before branching further into smaller and smaller bronchioles. The smallest of these are the terminal bronchioles which connect to the alveolar sacks by way of the alveolar ducts. (Centers for Disease Control & Prevention, part of the United States Department of Health & Human Services.)

as *lobes*. The primary bronchus splits into one secondary bronchus for each of these lobes. The right lung typically has three lobes, so the right branch splits into three secondary bronchi. The left lung typically has two lobes, so the left branch splits into two secondary bronchi. Once inside their respective lobes, the secondary bronchi split repeatedly forming up to 10 tertiary bronchi. Each of these tertiary bronchi subsequently split over and over until they form more than 6000 tiny tubes called *terminal bronchioles* each having a diameter of approximately 400 μm. At the end of each terminal bronchiole is a collection of small air sacks called *alveoli* all piled on the end of the tube like a lobe of cauliflower as shown in Figure 3.6. Each terminal bronchiole is connected to the alveolar sacks by way of alveolar ducts, each measuring about 100 μm in diameter. The sacks themselves are approximately 200 μm in diameter, and there are more than 100 million alveolar sacks in each lobe. Studies vary quite a bit on how many total alveoli are in both lungs combined, meaning all 5 lobes, with estimates ranging from 274 to 790 million and an overall average of about 480 million [4]. Studies are also diverse on the total alveolar surface area with numbers ranging from $70\ m^2$ to over $100\ m^2$. The total lung capacity, the maximum amount of air the lungs can hold without rupturing, ranges from $5\ L$ to $6\ L$. These numbers are interesting in contrast with the earlier reference on page 64 citing the NIH study (see footnote 2) indicating most adult humans only use about half a liter for each breath, or about 8% to 12% of our total lung capacity.

3.4   The Alveoli

Inside each alveolar sac is where the blood receives oxygen and surrenders carbon dioxide and other gases. The walls of each sack are incredibly thin, little more than a membrane as shown in Figure 3.7. The outer surface called the *capillary epithelium* is riddled with blood

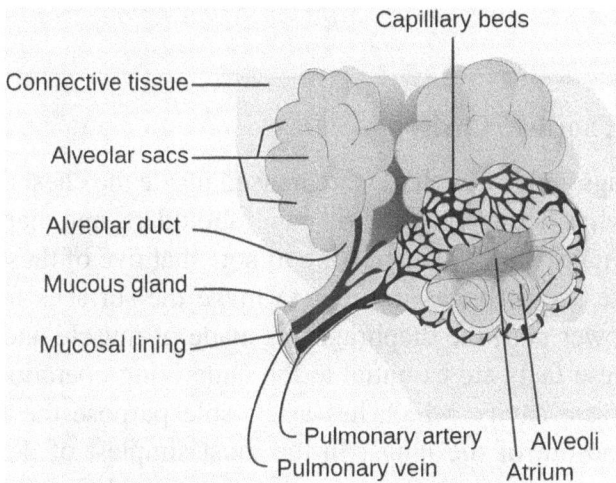

FIGURE 3.7
Inside the each alveolar sac is where the blood receives oxygen and surrenders carbon dioxide and other gases. The walls of each sack are incredibly thin, little more than a membrane. At this close proximity, both oxygen and carbon dioxide can easily move back and forth through the membrane facilitating the gas exchange.

vessels. The inner surface in contact with the air is called the *alveolar epithelium*. In between is the *fused basement membrane* which provides structure to the alveolar sack. The three surfaces combine to form the *respiratory membrane* which is less than 500 *nm* thick. At this close proximity, both oxygen and carbon dioxide can easily move back and forth through the membrane facilitating the gas exchange. The inner surface is lightly coated with a thin layer of surfactant which not only protects the alveolar surface from stray contaminants that managed to evade the filters, but it also helps keep the sack inflated even during exhale. At this small diameter, the surface tension in the water used to make the surfactant provides a tiny amount of force, enough to keep the sack in its roughly spherical shape. Without this surfactant, the sack would collapse on exhale and need to be forcibly reopened during inhalation. This would be a physically exhausting process requiring many times more work for each breath, and patients suffering from this condition quickly find themselves in respiratory distress, hence it is known as *respiratory distress syndrome*.

3.5   The Thoracic Cavity

The lungs and lower airway are housed inside the chest or *thoracic cavity* which is bounded by the rib cage, shoulders, and diaphragm, as shown in Figure 3.8. It is interesting to note that five of these six sides includes the use of bones in order to make the surfaces fairly rigid. Only the lower part, the diaphragm, is made of muscle and therefore pliable. These facts are essential to the underlying operational design of the thoracic cavity, which has as its sole purpose the movement of air in and out of the lungs. In the most simplest of descriptions, and therefore not entirely accurate, the five bone-laden sides form a rigid boundary, and the diaphragm acts like the bellows. In short, the diaphragm moves down and "sucks the air in," and then the diaphragm moves upward and "blows the air out." Although the finer details, which we will cover shortly, will somewhat augment this explanation, the fundamental physics of this loose description still holds.

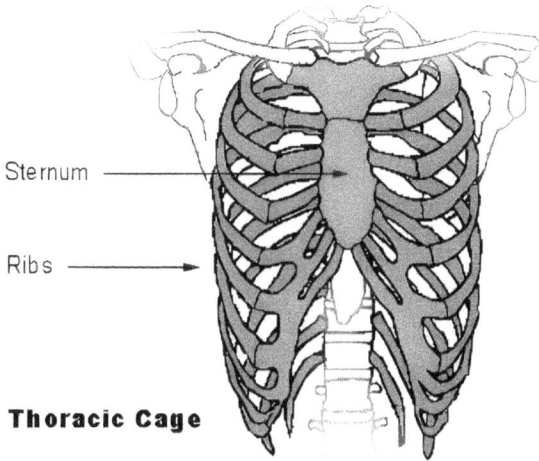

Sternum

Ribs

Thoracic Cage

FIGURE 3.8

The lungs and lower airway are housed inside the thoracic cavity which is bounded by the rib cage, shoulders, and diaphragm. It is interesting to note that five of these six sides includes the use of bones in order to make the surfaces fairly rigid. Only the lower part, the diaphragm, is made of muscle and therefore pliable. (Centers for Disease Control & Prevention, part of the United States Department of Health & Human Services.)

3.5.1   Chest Wall Construction

The most obvious part of the thoracic cavity is the chest wall, but in actuality this construction covers the front (anterior), back (posterior), and sides (lateral) parts of the thoracic chamber, since the rib cage literally surrounds the lungs. The chest walls on each of these four sides are made of a dozen pairs of bones, ribs, between which reside the *intercostal* muscles. On the outside of this foundation are the various layers of fat and skin tissues. On the inside is a thin membrane called the *parietal pleura* which coats the inside of the thoracic cavity and provides a smooth and slippery surface. The lungs are also surrounded by a membrane called the *visceral pleura,* and these two smooth and slippery surfaces are separated by a gap known as the *pleural space* which contains a lubricant known as *pleural fluid.* These membranes and lubricants are all to protect the lungs, keep them moist and pliable, and prevent any damage. Furthermore, filling the space with a fluid has certain mechanical advantages over a simple air gap, since fluids neither expand nor compress under changes of pressure, and when dealing with fluids, it is all about the pressures.

3.5.2 Breathing Mechanics

Under normal breathing conditions, the chest wall and diaphragm expand, increasing the volume of the thoracic cavity. Since the pleural fluid residing inside the pleural space cannot similarly expand, this generates a low pressure inside the alveoli and respiratory tract causing air to rush in and fill the increased volume. During exhalation, the process is reversed, the chest wall and diaphragm contract, reducing the thoracic volume. Again, the pleural fluid cannot similarly contract, so the pressure inside the alveoli and respiratory tract increases and drives the contained air volume out of the lungs.

The amount of force necessary to move the air in and out of the lungs is not confined the air movement, itself. The tissues of the chest wall and the lungs also have elastic resistance that must be overcome. There are also limits to how much pressure can be tolerated before the tissues lose their structural integrity and either rupture or collapse. The surface membranes of the alveoli, for example, have an elastic limit of approximately 20 *cm* H_2O or about 2000 *Pa*. In general, the change in lung volume for a given pressure change is known as *lung compliance,* and is typically measured in units of *mL/cm* H_2O. In other words, the increase (or decrease) in volume in units of *mL* for every 1 *cm* H_2O of pressure change. A normal human lung will see values of about

$$C_L = \frac{\Delta V}{\Delta P} = 200 \, \frac{mL}{cm \, H_2O} \tag{3.5}$$

As the definition suggests, higher values indicate lungs that are more easily distended (or collapsed); whereas, smaller values are indicative of "stiff lungs" which are not easily distended, thus indicating a situation where considerable effort is required for adequate air exchange. The chest wall, itself, also has an elastic content. Both the alveolar sacks and the chest wall have "natural" or relaxed positions that require no effort to maintain. Pressure is required to make them smaller or larger in volume. This is why you have to "work" to inhale, but most

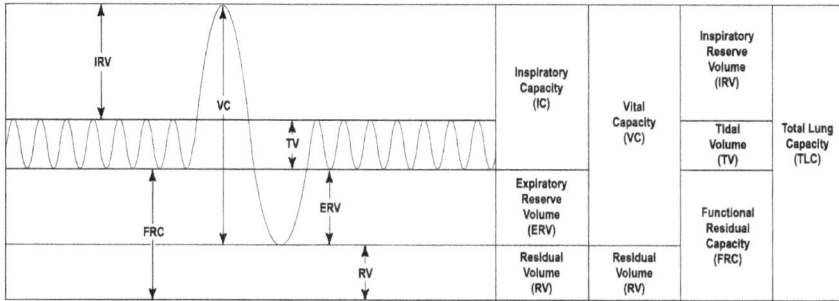

FIGURE 3.9

This chart shows the breakdown of the total lung capacity (TLC) of 6 *L* based on various stages of the respiratory cycle. Normal breathing in and out involves the Tidal Volume (TV) of about 500 *mL*, but we can manually increase this by an additional 3.1 *L* (IRV). Similarly, we can forcibly exhale an additional 1.2 *L* (ERV). However, the lungs will always attempt to retain the last 1.2 *L* (RV).

normal exhalation processes occur passively as the chest, diaphragm, and alveoli all return to their "natural" conditions.

3.5.3   Volumes and Capacities

Normal adult human lungs typically have a total lung capacity (TLC) of approximately 6 *L*. This quantity is the total holding capacity of the lungs, and most humans will never use even close to that amount. As mentioned previously on page 72, the average relaxed volume of air exchanged in each breath, or *tidal volume* (TV), is approximately 500 cm^3, or just half a liter as illustrated in Figure 3.9. Under heavy exertion, we can "breath heavily" and increase this amount another 3.1 *L* (IRV) to a total of about 3.6 *L* of air. At this point, the lungs are completely full. This number is less than the total lung capacity (TLC) because the remaining 2.4 *L* constitutes air that normally would never leave the lungs. In other words, the lungs are never really empty. They can't be. Should the lungs be completely squished down until all air has left, the interior surfaces, all covered in thick mucous, would stick together. Recall the combined interior surface area is nearly 100 m^2. The pressure required to pull these surfaces back apart would be astronomical. Consequently, the body always retains approximately

1.2 *L* in reserve, called the *residual volume* (RV). This air is distributed throughout the lungs in order to keep the surfaces apart. Furthermore, the "natural condition" mentioned previously for the chest, diaphragm, and alveoli leaves a combined volume of 2.4 *L*. You can push out the additional 1.2 *L* by forcibly exhaling (ERV), but you cannot (and do not want to) do anything about the remaining reserve volume. Combining the normal tidal volume (TV) with the deep breath (IRC) and forcible exhale (ERV) gives a total of 4.8 *L* which is called the *vital capacity* (VC). This amount includes everything except that residual volume kept in reserve in order to keep the lungs open and operating.

3.6   Gas Exchange Mechanics

As we discussed back in Section 2.1.6, the net average pressure of a gas is a time-averaged number of collisions per second against the container wall. That description is sufficient if you consider only a single type of gas molecule. However, if there are multiple species present in the gas, that time average will be a mixture of the collisions of all of the different species present; furthermore, the resulting pressure will be a combination, more specifically a *linear* combination, of all of the different species present in the gas sample. This is the basis for Dalton's Law of Partial Pressures.

The Earth's atmosphere is a composite of oxygen, nitrogen, water, carbon dioxide, argon, neon, krypton, hydrogen, helium, etc. so on and so forth. The major players are, of course, oxygen (20.946%), nitrogen (78.084%), argon (0.934%), and carbon dioxide (0.033%). The other atmospheric components, including solids and liquids, make up less than 0.003%. Since

$$P_{O_2} + P_{N_2} + P_{Ar} + P_{CO_2} + \cdots = 1 \ atm \qquad (3.6)$$

the *partial pressure* of oxygen will be roughly 0.20946 *atm*, for nitrogen 0.78084 *atm*, and so on.

Pressure is also a type of energy density, and in Section 2.1.13 it was shown that this energy density would even out over time such that the energy and population were both uniformly distributed throughout the available volume. This was tied back to the statistical thermody-namics principle of *equipartition,* or equal distribution, developed by Boltzmann in Section 2.1.10. In Section 2.1.13, this was shown to cause particle migration from regions of high population to low. When multiple species are present, this principle holds for each species inde-pendently until the same percent population is present throughout the volume, just like the raw numbers and energies throughout the volume.

Equilibrium can be loosely described as that situation where the conditions of a system do not change. That could be when all the forces acting on an object balance out such that the net force is 0 resulting in no acceleration. It could be when the heat transfer between two objects are equal in both directions resulting in no net gain nor loss by either object. In the case of pressures, equilibrium is established when the pressures at both locations are the same so the net number of collisions per second on both sides of the boundary are identical thereby generating no net force. In the case of mixed gases, such as atmospheric air, that extends to partial pressures as well as net pressure.

When blood arrives at the membrane of the alveoli sack, the oxygen content is slightly below that of the atmospheric air inside the sack. Conversely, the carbon dioxide levels in the blood are slightly higher than atmosphere. Therefore, a pressure gradient is created between the side with a higher population versus the side with the lower population. Since the membrane is very thin, the gas molecules are able to move from the high population side to the lower population side during the time interval in which the blood is in contact with the alveoli. In this way oxygen migrates from the air into the blood, and carbon dioxide is able to leave the blood and enter the air.

4

Physics of Pulmonary Functionality

As covered in Section 3.5.3, the lungs are required to move from a half-liter of air during quiescent periods to nearly five liters of air in critical situations several dozens of times every single minute. To do this, the thoracic cavity must provide the pressures necessary to initiate airflow in and out of the lungs in the required period of time. In this chapter we will examine the physical parameters necessary for normal lung activity under nominal conditions.

4.1   Upper Airway Flow Analysis

In Section 3.1.1 we showed that a volume of $500\,cm^3$ flowing uniformly through the $4.27\,cm^2$ area between the conchae requires a flow speed of $39.0\,cm/s$. This is a direct application of Equation 2.89.

$$(500\ cm^3)\left(\frac{20}{min}\right) = v\,(4.27\ cm^2) \tag{4.1}$$

$$v = \left(\frac{10\,000\ cm}{4.27\ min}\right) = 2342\ cm/min = 39.0\ cm/s \tag{4.2}$$

However, this assumes the entire $4.27\ cm^2$ cross-sectional area shown in Figure 4.1 is available for flow, and that is not true. The viscosity of air is small, but it is not 0, which means the boundary layer will similarly be small but non-zero. Equation 3.1 determined a Reynolds Number of 466 through this area which suggests a boundary layer of approximately

$$\frac{0.38\,(2.07\ cm)}{(466)^{\frac{1}{5}}} = 0.23\ cm \tag{4.3}$$

DOI: 10.1201/9781003683476-4

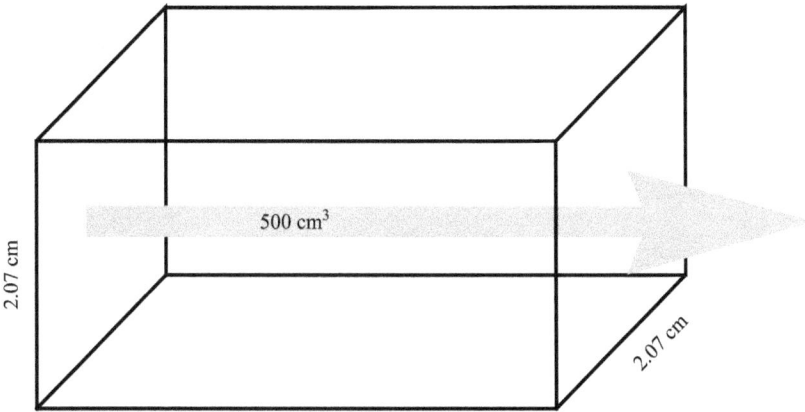

FIGURE 4.1
We first consider the flow through the nasal cavity without a boundary layer. The entire half-liter is allowed to pass through the passageway unimpeded. We assume for simplicity that the cavity is roughly square measuring 2.07 *cm* on a side for a total cross-sectional area of 4.27 *cm²*.

which means the actual cross section available for flow is only

$$\mathcal{A} = (2.07\ cm - 2\,(0.23\ cm))^2 = 2.58\ cm^2 \qquad (4.4)$$

which is only 60% of the original area as shown in Figure 4.2.[1] Since the boundary layers cannot contribute to the volumetric flow, that increases our mainline speed to

$$v = \left(\frac{10000\ cm}{2.58\ min}\right) = 3876\ cm/min = 64.6\ cm/s \qquad (4.5)$$

If we take this mainline velocity as the tangential velocity of a rotating vortex with diameter 2.30 *mm*, that gives a rotation rate of

$$\omega = \frac{v}{r} = \frac{646\ mm/s}{2.30\ mm} = 280.7\ s^{-1} \approx 2681\ rpm \qquad (4.6)$$

which is approximately the rotation rate of an industrial electric motor.[2]

---

[1]Note the fluid is bounded on all four sides, so we have accounted for a boundary layer on both surfaces of our hypothetically square area.

[2]Industrial electric motors run on A/C and typically have fixed RPM's of 3000 to 7200, which are multiples of the 60 cycle line frequency.

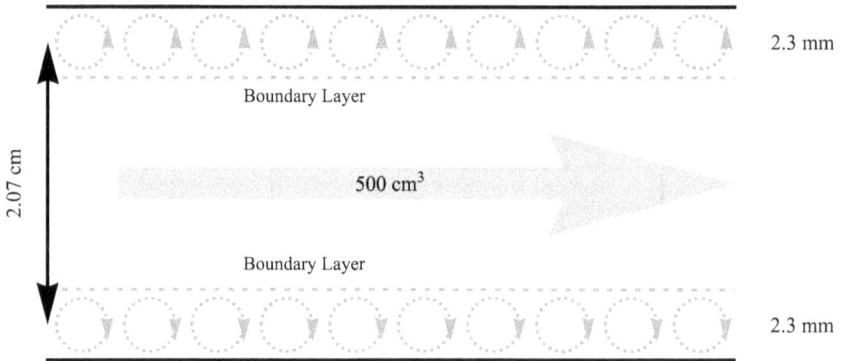

FIGURE 4.2
We now consider the flow through the nasal cavity where a 2.3 *mm* boundary layer has formed. The half-liter of air is now required to pass through a passageway of slightly reduced cross-sectional area. We assume for simplicity that the cavity is roughly square measuring 2.07 *cm* on a side as before, but the total cross-sectional area is now reduced to 2.58 *cm²*.

With this rotational velocity, the centrifugal driving force on any debris will be given by Equation 2.11

$$\frac{F_C}{m} = \frac{v^2}{r} = \frac{(64.6 \; cm/s)^2}{1.15 \; mm} = 362.8 \; \frac{N}{kg} \qquad (4.7)$$

An average pollen grain has a mass of approximately 20 *ng* and diameter of 100 *μm*. The centrifugal force acting on such a granule would be

$$F_C = \left(362.8 \; \frac{N}{kg}\right)(20 \times 10^{-12} \; kg) \approx 7.255 \; nN \qquad (4.8)$$

This doesn't sound like much, but bear in mind this is a single pollen grain, which is very small, so it doesn't take much force to make it move.

The settling velocity for a sphere through a viscous fluid is given by the ratio of the particle weight to the drag force by the fluid.

$$v_{set} = \frac{m \, g}{6\pi \, \eta \, r} \qquad (4.9)$$

where m is the buoyant weight of the sphere suspended in the media, g is the acceleration of gravity, r is the radius of the spherical particle, and η is the viscosity of the media. The pollen granule has an average density of

$$\rho_P = \frac{3\,m}{4\pi r^3} = 38.2\,\frac{kg}{m^3} \qquad (4.10)$$

which gives a buoyant mass of

$$m_B = \frac{4}{3}\pi\,r^3\,(\rho_P - \rho_{air}) = 1.948 \times 10^{-11}\,kg \approx 19.48\,ng \qquad (4.11)$$

so the settling velocity given by the centrifugal acceleration would be

$$v_{set} = \frac{\left(1.948 \times 10^{-11}\,kg\right)\left(362.8\,\frac{N}{kg}\right)}{6\pi\left(1.73 \times 10^{-5}\,\frac{kg}{m\,s}\right)(50\,\mu m)} = 43.34\,cm/s \qquad (4.12)$$

At this velocity, it would take the pollen granule 2.656 *ms* to travel the radius of the turbulent vortex, assuming the boundary layer was one vortex in diameter. In contrast, the vortex churning at 280.7 *rad/s* would take 22.38 *ms* to complete a single rotation, so the pollen granule would be driven outward to the wall of the nasal passage in less than one-tenth of a rotation. The arched and irregular curvature of the conchae causes the much heavier dust and pollen particles to enter the turbulent boundary regions where the centrifugal force quickly drives them into the thick mucous coated regions and become trapped.

At this rotational velocity, contaminants must be driven to the wall in less than half a rotation to avoid the possibility of re-entering the main flow. That requires a minimum settling velocity of 10.28 *cm/s*. If we rewrite Equation 4.9 in terms of density we get

$$v_{set} = \frac{2\,\rho\,a_c\,r^2}{9\,\eta} \qquad (4.13)$$

Using our minimum settling speed, that gives the relation

$$\rho\,r^2 = 2.207 \times 10^{-8}\,\frac{kg}{m} \qquad (4.14)$$

If we assume a density comparable to our pollen granule, correlating Equation 4.14 with Equation 4.9 suggests a maximum diameter of 34 μm. Similarly, if we assume a particle of equal size to our pollen granule, that would give a minimum density of 4.413 kg/m^3 which equates to a mass of 2.311 ng.

Pulling the air through the confined and intentionally turbulent nasal passages, and creating all of these swirling vortices, requires a considerable amount of energy. This results in a potential energy loss in the form of a pressure drop between the atmosphere and back of the upper airway. A flow of 10000 cm^3/min is equivalent to 0.0100 m^3/min. The nasal passageway is typically about 10 cm in length with a total volume of 25 cm^3 which gives an average 2.5 cm^2 of cross-sectional area of which perhaps half is occluded by the conchae. Assuming a roughly square pipe, that would be an average 1.58 cm width and height. Accounting for the reduction caused by the boundary layers, this reduces our effective pipe size to $d = 11.2$ mm. Using Equation 2.83 with atmospheric pressure as the driving mechanism we get an estimated

$$\Delta P = \left(7.57 \times 10^4\right) \frac{(0.090 \ m) \left(0.0100 \ \frac{m^3}{min}\right)^{1.85}}{(11.2 \ mm)^5 \left(1.03 \ \frac{kg}{cm^2}\right)} \quad (4.15)$$

$$\Delta P = 7.49 \times 10^{-6} \frac{kg}{cm^2} \approx 734 \ mPa \quad (4.16)$$

or nearly three-quarters of a *Pascal* in pressure, which is why it is slightly more difficult to breathe through your nose than mouth.

4.2 Larynx Analysis

Passage of air into the larynx poses two problems. The air must first execute a turn into the laryngopharynx followed by passage through the vocal folds and past the ligaments of the larynx, itself. The angle between the oral cavity to the larynx is roughly 90° whereas the angle from the oral cavity to the pharynx roughly 20°. That makes the turn from the pharynx to the larynx roughly 70°. The inner dimensions of

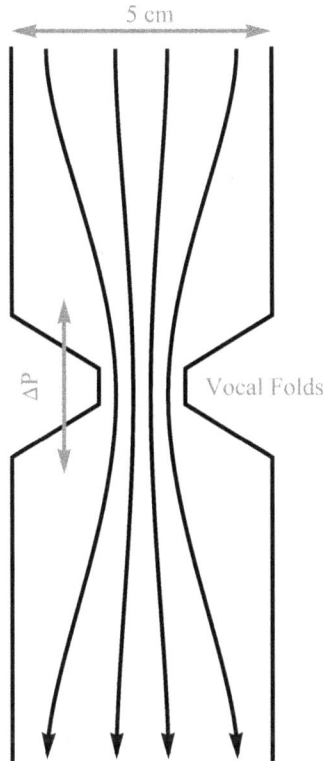

FIGURE 4.3

The inner dimensions of the larynx box is approximately cubic measuring roughly 5 *cm* and about a centimeter thinner anterior-to-posterior than left-to-right. Occluding that flow area are the vocal folds.

the larynx box are approximately cubic measuring roughly 5 *cm* and about a centimeter thinner anterior-to-posterior than left-to-right. That gives the larynx a total of about 100 *cm*^{3} in volume and a maximum flow area of 20 *cm*^{2}. Occluding that flow area are the vocal folds as shown in Figure 4.3. In a relatively healthy adult, maximum openings can range from 32° to 77° [6] and form an approximate isosceles triangle with a height equal to the inner diameter of the larynx cavity.

When air faces a constriction such as a nozzle or sudden narrowing, air will accumulate in the upstream region generating the pressure elevation necessary to force the air to both compress and accelerate as it moves through the nozzle. Equilibrium is achieved when the required mass flow rate balances the input. The correlation can be written as

$$v^2 = c \frac{DP}{L} \qquad (4.17)$$

where v is the linear speed of the air through the nozzle, D is the diameter of the nozzle, L is the length of the nozzle, and P is the upstream pressure driving the air through the nozzle. The constant c is empirically determined to be the following for air:

$$c = \left(\frac{3125}{18}\right)\left(\frac{ft^4}{s^2 \, ozf}\right) = 5.38977 \left(\frac{m^2}{Pa \, s^2}\right) \qquad (4.18)$$

The velocity v is regulated by the required volumetric flow rate and the corresponding area between the vocal folds, so the pressure difference required to initiate that flow can easily be found.

$$P = \frac{L v^2}{c D} \qquad (4.19)$$

Vocal folds can be up to 5 *mm* in thickness. If we assume the vocal folds occlude two-thirds of the airway, then a 500 *mL* breath taken 20/*min* would require a flow velocity of

$$(500 \, cm^3)\left(\frac{20}{min}\right) = v \left(\frac{1}{3}\right)(20 \, cm^2) \qquad (4.20)$$

$$v = 25 \, cm/s \qquad (4.21)$$

which would then require a pressure drop of

$$P = \frac{(5 \, mm)(25 \, cm/s)^2}{\left(5.38977 \, \frac{m^2}{Pa \, s^2}\right)(2.582 \, cm)} = 2.25 \, mPa \qquad (4.22)$$

Sound production requires the vocal folds to come together and vibrate as the combination of Venturi Effect and pressure differential cause them to be pushed apart and then drawn back together. The frequency of this vibration is driven by the tension in the vocal folds generated by the vocalis muscles much like the vibrations on a stringed instrument. The amplitude of the sound is regulated by the driving pressure generated from the lower airway and thoracic cavity. Only those frequencies which resonate in the larynx, pharynx, and oral cavities will generate sufficient volumes, so the specific set of frequencies available are dictated by the physical dimensions of the respective cavities. This selection is what makes each person's voice unique.

Using our typical larynx dimensions of $5\ cm \times 5\ cm \times 4\ cm$ from earlier, and assuming the vocal folds are exactly centered vertically in the larynx chamber, our fundamental or resonant frequencies become

$$f_0 = \frac{v_s}{2D} = \frac{340\ m/s}{2(5\ cm)} = 3400\ Hz \tag{4.23}$$

for the width of the larynx chamber, and

$$f_0 = \frac{v_s}{2D} = \frac{340\ m/s}{2(4\ cm)} = 4250\ Hz \tag{4.24}$$

for the depth of the larynx chamber. The height of the chamber should extend from the larynx across the width of the pharynx, since the turn is close to a right angle, and in this case the vocal folds would be one end of the resonance cavity. That makes the effective length nearly $6\ cm$ for a fundamental frequency of

$$f_0 = \frac{v_s}{2D} = \frac{340\ m/s}{4(6\ cm)} = 1417\ Hz \tag{4.25}$$

Resonance also occurs within length, width, and depth of the oral cavity. Examining these dimensions leads to a whole new set of frequencies. The oral cavity is typically about 8 *cm* in length which would yield a lengthwise fundamental frequency of about 2125 *Hz*. If we examine the length from the larynx to the nasal cavity, which measures roughly 18 *cm* for males and 16 *cm* for females, we get frequencies of 472 *Hz* and 531 *Hz*, respectively.[3]

These are but a few of the resonance conditions possible. There are many, many more as one considers the different possibilities, including cross diagonals. Each of these situations can and do occur, which give a person's voice unique and quite complex tonal qualities. In various branches of acoustics and music theory, these fundamental frequencies are also known as *formants* since these dimensions are relied upon to *form* the various notes and harmonics used in singing and speaking.

_____

[3]Note that the vocal cords again serve as one end of the resonance chamber.

4.3 Trachea Flow Analysis

The inside diameter of the trachea is typically about 19.5 *mm* for males and 17.5 *mm* for females and has a length of 10 *cm* to 16 *cm* between the bottom of the larynx to the bronchial branch or bifurcation. For a 19 *mm* pipe, a 500 *mL* breath taken 20/*min* would require a mainline flow velocity of

$$v = \frac{(500 \ cm^3)\left(\frac{20}{min}\right)}{\frac{\pi}{4}(1.90 \ cm)^2} = 3527 \ cm/min = 58.78 \ cm/s \qquad (4.26)$$

which yields a Reynolds Number of

$$Re = \frac{(1 \ kg/m^3)(14.7 \ cm/s)(1.90 \ cm)}{1.73 \times 10^{-5} \ \frac{kg}{m \ s}} \approx 646 \qquad (4.27)$$

which could be laminar. At the very least, the degree of turbulence in this flow is small, which is why you are asked to take a "deep breath" when doctors listen to your breathing. Taking this "deep breath" increases the tidal volume to about 3 *L* per breath which increases the mainstream flow velocity to 353 *cm/s* and the Reynolds Number to more than 3874. Since the incoming air has already passed over the obstruction caused by the vocal folds, this flow very likely to be quite turbulent and definitely audible.[4] With a Reynolds Number of 646, the estimated boundary layer is only 1 *mm* thick which reduces the effective diameter to 17 *mm*, increasing the mainstream flow velocity to 73 *cm/s*.

With a thin boundary layer and nominal flow velocities, the viscous drag due to the walls of the trachea as given by Equation 2.83 are appreciably small

$$\Delta P = (7.57 \times 10^4) \frac{(0.140 \ m)\left(0.0100 \ \frac{m^3}{min}\right)^{1.85}}{(19.0 \ mm)^5 \left(1.03 \ \frac{kg}{cm^2}\right)} \qquad (4.28)$$

$$\Delta P = 8.29 \times 10^{-7} \ \frac{kg}{cm^2} \approx 81.3 \ mPa \qquad (4.29)$$

---

[4]As any submariner can tell you, only turbulence can be heard. Laminar flow is silent even to the most sensitive listening gear.

The pressure required to drive the flow through the trachea can be determined by solving Equation 2.89 for the pressure

$$\Delta P = \frac{8 L Q \eta}{\pi R^4} \tag{4.30}$$

For the flow currently under consideration, that would require a pressure difference of only

$$\Delta P = \frac{8 \, (0.140 \, m) \left(\frac{m^3}{6000 \, s}\right) \left(1.73 \times 10^{-5} \, \frac{kg}{m \, s}\right)}{\pi \, (0.0190 \, m)^4} = 7.89 \, mPa \tag{4.31}$$

in addition to the pressure required to overcome the viscous drag for a total of 89.2 *mPa*.

4.4   Terminal Bronchioles Analysis

Once the flow reaches the bronchial tree, the flow separates into five parts, one for each lobe. Each of these five parts then branches into 6000 additional parts for a total of 30,000 individual tubes; however, the collective volumetric flow rate must be constant according to continuity. Therefore, the flow rate in each of these tubes must be

$$q = \frac{(500 \, mL) \left(\frac{20}{min}\right)}{6000} = \frac{5 \, cm^3}{3 \, min} \tag{4.32}$$

These tubes are, of course, the terminal bronchioles each of which feed a small collection of alveoli. They have an average diameter of 400 μm, so the mainline velocity must be

$$v = \frac{\frac{5 \, cm^3}{3 \, min}}{\frac{\pi}{4} (400 \mu m)^2} = 1326 \, cm/min = 22.1 \, cm/s \tag{4.33}$$

which produces a Reynolds Number of

$$Re = \frac{(1 \, kg/m^3)(22.1 \, cm/s)(400 \mu m)}{1.73 \times 10^{-5} \, \frac{kg}{m \, s}} \approx 5.11 \tag{4.34}$$

which is clearly laminar, as it should be. At this small of a diameter, the flow can ill afford any kind of boundary layer development, and it is unlikely that the alveoli would respond well to any turbulence.

Again using Equation 2.83, the viscous pressure drop over the approximate 1 *mm* length of the terminal bronchioles is approximately

$$\Delta P = (7.57 \times 10^4) \frac{(0.001 \ m) \left(1.67 \times 10^{-6} \ \frac{m^3}{min}\right)^{1.85}}{(0.400 \ mm)^5 \left(1.03 \ \frac{kg}{cm^2}\right)} \quad (4.35)$$

$$\Delta P = 1.47 \times 10^{-7} \ \frac{kg}{cm^2} \approx 14.4 \ mPa \quad (4.36)$$

The pressure difference required to drive this flow is

$$\Delta P = \frac{8 \ (0.001 \ m) \left(\frac{5}{180} \ \frac{cm^3}{s}\right) \left(1.73 \times 10^{-5} \ \frac{kg}{m \ s}\right)}{\pi \ (200 \times 10^{-6} \ m)^4} = 765 \ mPa \quad (4.37)$$

For a total required pressure difference of approximately 780 *mPa*.

4.5   Required Respiratory Driving Pressures

The pressures required to move the air in and out of the lungs through the airways would be the nominal required driving pressure plus any viscous losses due to pipe lengths and obstructions. That includes the nasal passages, laryngopharynx, pharynx-to-larynx turn, larynx and vocal cords, trachea, bronchial tree, and terminal bronchioles. The pressure drop determined for each of these items is listed in Table 4.1 and combine for a total of 1616 *mPa* which is the equivalent to a 0.0165 *cm* column of water (*cm* H_2O); however, normal lung resistance should be roughly 0.5 *cm* H_2O and normal ventilation pressures are from 3 *cm* H_2O to 5 *cm* H_2O (see footnote 4).

This analysis thus far is not flawed, just incomplete. To this point we have only considered only the motion of the fluid itself and possibly its interactions with the boundaries. What we have not considered are the boundary walls themselves, save for the extent to which the moving fluid would interact with or be constrained by that boundary. We have not considered the airway or the lungs themselves.

TABLE 4.1

A list of all itemized pressure drops for the entire airway and the total pressure drop that is accumulated in the movement of the air

Airway Component	Pressure Drop (*mPa*)
Nasal cavity	734.
Pharynx length of 18 *cm*	0.828
Pharynx to larynx 70° turn	4.6
Larynx pressure drop	2.246
Trachea drag and drive	89.2
Bronchiole pressures	780.
Total	1616.0

4.5.1 What About Compressibility

Throughout the analysis we have used empirical solutions for compressed air flow under appropriate circumstances; however, we have not looked specifically just how much the air has compressed at any time. Before moving on to the mechanical aspects of respiration, I would like to take a moment and finish off the discussion about air as a compressible gas.

Normal atmospheric pressure is 1 *atm* which is 101,325 *Pa* or 1033.93 *cm* H_2O. The air all around you at any given moment is at roughly this pressure. When you inhale, your lungs reduce their internal pressure by a small amount. The room pressure compensates by driving the relatively high pressure outside air into the low pressure region through the respiratory canal previously described. When you exhale, the process is reversed and the higher pressure air in your lungs is vented to the relatively lower pressure in the room.

According to Table 4.1, the pressure difference required to move air in and out is 1.616 *Pa*. Using the bulk modulus for air of 101 *kPa* indicated in Section 2.1.7.1, Equation 2.26 gives a volume change of

$$\frac{\partial V}{V} = \frac{1.616 \; Pa}{-101 \times 10^3 \; Pa} = -1.600 \times 10^{-5} \tag{4.38}$$

so a $500\ cm^3$ tidal volume of air would change by $0.008\ cm^3$ changing the air density from $1.000000\ kg/m^3$ to $1.000016\ kg/m^3$.

If we take the standard $5\ cm\ H_2O$, this equates to $490\ Pa$ and gives a volume compression of

$$\frac{\partial V}{V} = \frac{490\ Pa}{-101 \times 10^3\ Pa} = -4.852 \times 10^{-3} \qquad (4.39)$$

so a $500\ cm^3$ tidal volume of air would shrink to $498\ cm^3$ changing the air density from $1.00000\ kg/m^3$ to $1.00488\ kg/m^3$.

In either case, I think it is safe to assume that, even at maximum ventilator pressures, the change in air density is small and can safely be neglected. Our focus should be on the dynamics of the compressible fluid, not on the density.

4.6 Elastic Properties of the Lungs

One common phrase found throughout medical literature is the term, "increase in lung volume." Unlike most mechanical devices, the lungs do not have a fixed volume. Inhalation literally increases the volume of the lungs, and exhalation correspondingly decreases that lung volume. Of particular importance are the two boundary layers that must change in order for the total internal volume of the lungs to change: the boundaries of the lungs (alveoli) and the bounding surface of the thoracic cavity (chest wall). Quite literally, filling the alveoli requires that these tiny sacks get bigger, and simultaneously the chest wall must move out of the way to make room for the larger lungs. Both of these actions will require additional air pressure if instigated externally, such as with a ventilator or ventilation bag. In the case of spontaneous respiration, the chest muscles and diaphragm will produce the pressure required to move the air and expand or contract the lungs, so any external pressure required to move these muscles is no longer necessary.

A recent study in 2020 by Jawde et al. [1] found the average elastic modulus for the walls of the alveoli were approximately $5\ kPa$. The elastic modulus is another name for Young's Modulus which was defined by Equation 2.31 back in Section 2.1.7.3. The study looked at the results of the tensile stress on the alveolar wall causing the wall to stretch, thus the "sphere" to get larger.

The alveoli are said to have an average diameter 400 μm and thus a circumference of approximately 1257 μm. It is unclear in the literature if this is a maximum, minimum, or median value. To approximate, if we assume a total of 100 million alveoli in each of the 5 lobes, the required alveolar volume for a total lung volume of 2.4 L would be

$$V_{alv} = \frac{2.4 \, L}{500 \times 10^6} = 0.0048 \, \mu L \qquad (4.40)$$

which yields an average diameter of 209 μm. On the other hand, adding an additional 500 mL for a total of 2.9 L increases the alveolar diameter to 223 μm. That would require an increase of

$$\sigma = (5 \, kPa) \left(\frac{223 - 209}{209} \right) = 335 \, Pa \approx 3.4 \, cm \, H_2O \qquad (4.41)$$

which is consistent with common ventilator practice which states an increase of 5 $cm \, H_2O$ should increase lung volume by 500 mL[5]. The remaining pressure difference between the findings thus far and accepted values can be resolved by considering the one remaining piece: the chest wall.

Unfortunately, the chest wall is not as straight forward as the alveolar walls. The elasticity of the chest wall changes depending on the operational region from forced expiration to deep inhalation. If we confine ourselves to the "normal" range as in the previous example, the chest behaves very linearly with an expansion coefficient of approximately 2.6 $cm \, H_2O/L$.[6] For a 500 mL flow, this result in an addition 1.3 $cm \, H_2O$ of pressure required, bringing our total to 4.7 $cm \, H_2O$.

4.7 Final Thoughts

It is interesting to note that almost the entire amount of respiratory driving pressure required is due to the various elastic properties of the

---

[5]Ventilator management.
[6]Elastic properties of the respiratory system.

respiratory tissues, not the movement of the air. In fact, the pressures required for actual air flow through even the smallest tubules or most obstructed regions remains negligible. Indeed, most of the pressure requirements seem to be in the actual expansion of the lung tissues, themselves, accounting for more than 68% of the total amount of pressure required.

Since the lung tissues have the largest impact on the ability to breath, and their ability to function is essential to the entire operation of the respiratory system, it stands to reason that any negative impacts on the lung tissues would severely, and possibly catastrophically, impact or negate their operational readiness. In the following and final chapter, I will explore the changes in the physics of respiration when the lungs or their connective airways are compromised by disease or physical damage.

5

Pulmonary Diseases, Damage, and Implications

Respiratory issues can be broken down into three major categories: those that restrict or impede the airways, those that restrict or impede the ability to generate the required pressures, or those that restrict or impede the ability of the lungs to exchange gases properly (or at all). In this final chapter, I will explore several common conditions and their impact on the physics of respiration in order to further understand why these conditions can cause varying difficulties ranging from annoying to life-threatening.

5.1 Nasal Congestion

Probably the most common human condition when it comes to breathing is the dreaded seasonal allergy, or worse yet, the "common cold." As horrible as it sounds, and I am not trying to make light of such a horrible feeling, but if your *only* problem is nasal congestion, you are much better off. Unfortunately, far too many common colds end up causing pneumonia, which I will discuss in the next section, in addition to the nasal congestion found in common allergies.

Nasal congestion is often associated with mucus either running out of the nose or down into the oral pharynx, known commonly as the *runny nose* or *postnasal drip,* sneezing, facial pain and redness, watery eyes, and possibly a headache. For some people, the pain can also spread to the ears. The difficulties in coping with the effects of this congestion will often leave one tired and irritable. The drainage from the sinus membranes can also create a bad taste in your mouth and possibly bad breath.

DOI: 10.1201/9781003683476-5

As discussed in Section 3.1.1, the nasal passageways serve two purposes in addition to just letting air in and out. One purpose is to humidify the air to nearly 100% saturation, and the other is filtration. This last part is where the seasonal allergies come into play, as the pollen that is removed from the air ends up irritating the nasal membrane. Pollen is a horrible thing surrounded by pointy spears that can cause physical damage to any membranes in which they come into contact. In addition to damage control responses, some pollen can also generate an allergic response which kicks the immune system into action. This is known as the *histamine* response and can vary from mild to life-threatening. For most people, the histamine reaction results in sneezing, runny nose, and watery eyes. Sneezing is the body trying to expel this irritant by forced exhalation through the nasal passageways. The runny nose is the body trying to both expel the irritant through irrigation and also trying to reduce the amount of irritation by lubricating the compromised areas. The watery eyes are due to a condition known as *rhinoconjunctivitis* which indicates the connection between the nasal (rhino) and eye (conjunctive) parts. When the body kicks in the histamine response, it cannot restrict the response to just the nasal area. The histamine related chemicals are in the blood, so they effect the eyes, nose, and throat simultaneously, thus your eyes feel the effects along with these other areas. The inflammation of the sinus membranes also causes the blood vessels in this area to enlarge in order to provide the region with the requisite materials to combat the infection or irritation. In some cases, this may lead to nosebleeds or bloody mucus.

The major respiratory issue with nasal congestion, be it due to allergies or colds, is the restriction of airflow. The conchae are already restricting flow through the nasal passages in an effort to filter out the air before it gets into your lungs. Allergies and colds can cause these conchae, and the other membranes, to swell which occludes the passageway even further. In severe cases, it can block the passages completely and, should the swelling continue, cause severe headaches and other medical problems.

As the membranes swell, the available cross-section decreases. This restriction in the airflow results in a higher pressure demand in order to meet the volumetric flow rates required as can be shown by the pressure solution to Equation 2.89 if written in terms of area $\mathcal{A}$

$$\Delta P = \frac{8\,L\,Q\,\eta}{\pi\,R^4} = \frac{8\pi\,L\,Q\,\eta}{\mathcal{A}^2} \tag{5.1}$$

This clearly demonstrates that decreasing the area by half will, at a minimum, increase the pressure demands by four. I say at minimum because we have not considered any increase due to changes in the viscous drag or increased turbulence that might also result. Once you can no longer meet the pressure demands, volumetric flow suffers and you start getting short of breath. In most cases, this results in switching over to "mouth breathing."

There is an age-old proverb that claims "there is no cure for the common cold," and they are largely correct. For one thing, the "common cold" is not singular. There are more than 200 different viruses that produce symptoms commonly associated with a cold, and the symptoms associated with a particular virus may vary slightly from person to person. Fortunately, the body is able to fight off each contraction of the disease within a few days. As Figure 5.1 demonstrates, most patients recover from whatever symptoms they might be experiencing within 10 to 15 days.

The most common viruses associated with a cold are in a family known as *rhinoviruses* from the root word *rhino* meaning "nose." However, other viruses in other families such as *parainfluenza, adenoviruses, enteroviruses,* and the human *metapneumovirus* can also produce similar symptoms.

The common cold cannot be cured, so treatment often resorts to addressing the symptoms. Antihistimines reduce the body's sensitivity to irritants like pollen, thereby reducing the reaction to those irritants including the swelling and mucus production. Anti-inflammatory and decongestants attempt to reduce the swelling of the membranes which should widen the nasal passages and allow air movement through the conchae again. This would also reduce any sinus pressure and the associated headache and facial pain that it might cause.

There are two major problems preventing us from "curing" the common cold. First, these viruses mutate rapidly and come in an extremely large variety of strains. From the virus point of view, this is necessary for survival, since the body learns how to combat the viruses

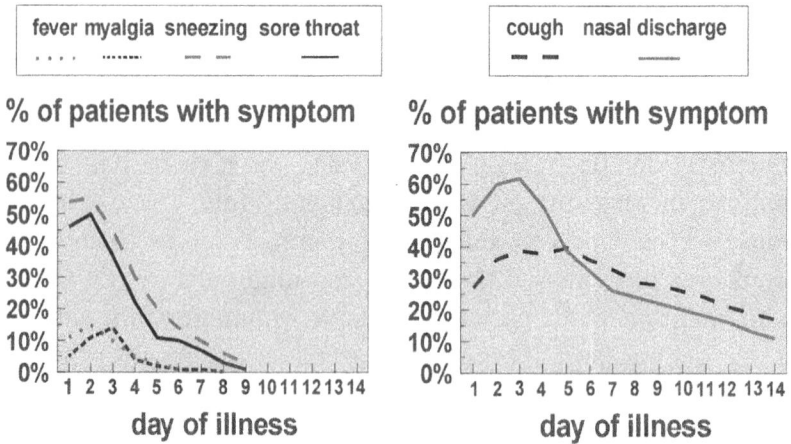

FIGURE 5.1

The most common symptoms of the common cold may include coughing, sneezing, runny nose, fever, sore throat, body aches, and watery
eyes. As these charts indicate, most patients recover within 10 to 15
days. (Centers for Disease Control & Prevention, part of the United
States Department of Health & Human Services.)

one strain at a time. Unfortunately, the immune response for strain A
will not work on strain B, so when the next one shows up the body
has to start all over again. This is partly why you can get cold after
cold after cold. The other reason is that it might not even be the same
family of virus. So, if you put it together, we have a large number of
variations within a family and a number of different families, so you
never know which one will come next, and one treatment might not
work on another family or variant.

The other problem is that the cold is caused by a virus, not a
bacteria. While both are living things, bacteria are cells in their own
right and can exist independently. Viruses are not a complete living
cell nor are they self-sustaining. They require a living host to inhabit,
so they actually reside inside a living cell. While antibiotics are very
effective in killing and preventing the growth of bacteria, often a whole
family of bacteria in fact, antivirals simply attempt to stop the virus
from multiplying. Unfortunately, they cannot "kill" the virus directly.
While popular opinion is that antivirals can be quite effective, passive

observation might disagree because of this subtle difference. With an antibiotic, the bacteria dies quite rapidly, and the patient receives relief in a matter of hours or a few days; however, the antiviral simply stops or inhibits replication. It still remains in the body, and the body will still ultimately have to deal with it, and that takes time, perhaps as long as a week or month.

5.2 Pneumonia

One of the most dangerous aspects of colds and the flu is the significant probability that it will result in "contracting" pneumonia. There are several kinds of pneumonia, each identified by their root cause, including direct aspiration of solids or liquids such as food or drink into the lungs and breathing in dangerous and damaging chemical substances such as gasoline or chlorine.

Bacterial pneumonia is, obviously, caused by some kind of bacteria. Common types of bacteria that cause pneumonia include Haemophilus influenzae, streptococcus, and Legionella, but there are many others. The bacteria is usually inhaled along with some host media. Legionella and influenza, for example, are usually inhaled while they are residing inside tiny contaminated water droplets, but bacteria, unlike a virus, does not necessarily require water to survive. Dry dust can serve as an equally good medium for many bacteria.

Viral pneumonia is, obviously, caused by some kind of virus. Most of these viruses are variants of influenza, such as influenza A, parainfluenza, and so on; however, other viruses such as the rhinovirus, respiratory syncytial virus (RSV) and adenoviruses can also cause pneumonia, particularly in small children. Viruses are living cells and require water to survive; therefore, most viruses are contracted by either inhaling contaminated water droplets or otherwise getting some kind of contaminated fluid into direct contact with your own blood.

Fungal pneumonia occurs when your lungs turn into a mushroom farm and are medically categorized as *pulmonary fungal infections*. The most common types of fungi include histoplasma which can exist as a form of mold or yeast that are supposed to live in soil contaminated by bird or bat droppings. Coccidioides immitis is a type of fungus which prefers dry desert areas. Cryptococcus neoformans is a type of yeast that also lives in soil contaminated by bird droppings. There are a great many more fungi out there that cause all kinds of problems, but they all have one thing in common: fungal infections are contracted by inhalation of the fungal spores. These spores are the "seeds" of a fungus plant and can take root in your respiratory system. Once they do, they are very hard to eradicate completely.

Parasitic pneumonia occurs when some kind of living organism takes up residence in your lungs. These might include paragonimus, ascaris, malaria, threadworm, dirofilariasis or some other parasite. Such infestations are very rare, and the ability to treat them varies from one species to another. Some infestations, such as paragonimus, are contracted by eating raw or under-cooked seafood. Others such as roundworms or threadworms are often contracted elsewhere in the body and migrate to the lungs.

No matter the type or cause, the consequences of pneumonia are very nearly the same. The tissues of the lungs, particularly in the alveolar sacks, become inflamed and could secrete fluid into the pulmonary space that is supposed to be used by the air, as shown in Figure 5.2. Once this space is filled by the liquid, the air no longer has access to that region and cannot reach the capillaries and exchange oxygen for carbon dioxide. As a result, lung efficiency is reduced, possibly to the point where an insufficient amount of oxygen can be provided to the blood for the patient to survive.

The most obvious treatment option would be to address the root cause of the problem, such as killing the bacteria, fungus, worms, or whatever. Unfortunately, this approach does not work as well with viruses, as antivirals are at best only moderately effective and at worst unavailable, and other sources cannot be as easily addressed. In some cases, the cause is not actually known, so one resorts to treating the symptoms. In that case, the main objectives would be to (1) increase the available lung capacity and (2) the efficiency of the current volume that is available for use.

Artificially increasing the respiratory pressure might reverse the osmotic process and drive the liquid secretions back into the blood, freeing up critical lung volume for gas exchange, but excessive

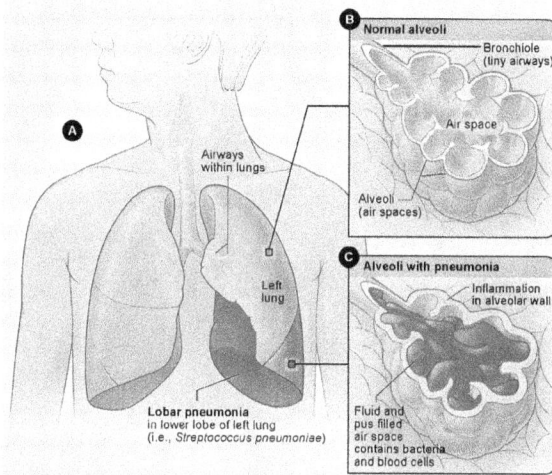

FIGURE 5.2
When a patient contracts pneumonia, the tissues of the lungs, particularly in the alveolar sacks, become inflamed and could secrete fluid into the pulmonary space that is supposed to be used by the air. Once this space is filled by the liquid, the air no longer has access to that region and cannot reach the capillaries and exchange oxygen for carbon dioxide. (National Institutes of Health, part of the United States Department of Health & Human Services.)

pressures can hyperextend and possibly damage the lungs by rupturing the alveolar sacks. Diuretics might also help the body extract some of that fluid and open up more lung volume, but this carries the risk of damaging other organs.

One can also enrich the oxygen content of the air to provide more oxygen to the space that is currently available, but this obviously has the firm upper limit of 100% pure oxygen. Cranking up the partial pressure in Equation 3.6 can help, but it still takes time.

5.3 Bronchitis

Just as pneumonia causes swelling and secretions in the alveolar sacks, and allergies can cause swelling and secretions in the nasal cavity,

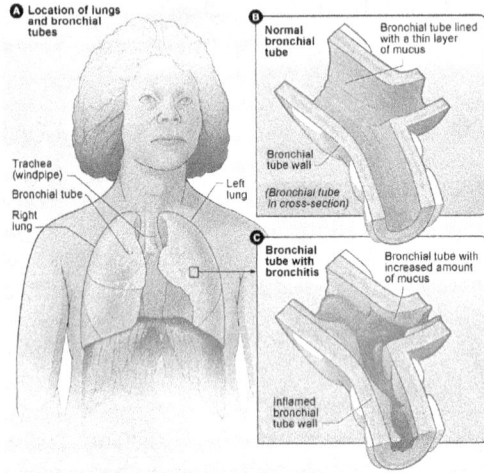

FIGURE 5.3
Bronchitis is the situation where the swelling and secretions occur in the bronchial tree, anywhere or everywhere from the branch down to the terminal bronchioles. As a result, the cilia are damaged or destroyed all along the effected passageways. (National Heart Lung & Blood Institute, part of the United States Department of Health & Human Services.)

bronchitis is the situation where the swelling and secretions occur in the bronchial tree, anywhere or everywhere from the branch down to the terminal bronchioles, as illustrated in Figure 5.3. As a result, the cilia are damaged or destroyed all along the effected passageways. The causes and some symptoms of bronchitis might mimic those of both nasal allergies and pneumonia and can even accompany or follow after these other issues.

The most prominent symptom of bronchitis is a pronounced cough which may or may not be productive, meaning you may or may not cough something up. Eventually, as the condition progresses, mucus will be created in the airway and ultimately expelled during the cough. The obstruction in the airway increases the turbulence and produces a noticeable wheezing or whistling as you breathe. The constant coughing might lead to a sore throat or a headache, or both. The struggles with breathing lead to chest pains and shortness of breath. If the cause is due to an infection, there might be a low grade fever.

If the cause is due to illness or immediate short term exposure to some kind of irritant, the onset can be quite rapid, a situation called *acute* in the medical profession. Like the cold itself, the worst of the symptoms pass after a few weeks as the whole thing clears itself up. In many cases, the person may not even be aware they had bronchitis, just attributing their "chest congestion" to being part of the cold. The most common cause of acute bronchitis is a viral infection, such as influenza or a rhinovirus.

Long term exposure to harsh or damaging chemicals or other irritating substances can cause long lasting or even permanent damage to the bronchioles causing the bronchitis to also become permanent, or *chronic*. This never-ending form of bronchitis is one of the three members of *chronic obstructive pulmonary disease, or COPD*. Although it is most commonly and publicly associated with smoking, any environment that presents long term exposure to harsh chemicals or otherwise contaminated air will suffice, such as those produced by cleaning and laboratory solutions, most notably chlorine or bleach, or toxic dust from industrial manufacturing processes such as fiberglass or sawdust.

As with the previous cases, the functional issue here is a narrowing of the airways due to inflammation and mucus buildup. This reduces the available passageway for the air and increases the amount of pressure accordingly. The deeper bronchioles have increasingly smaller areas and, between the swelling and thick mucus, can easily be obstructed thereby cutting off access to entire collections of alveoli. The increase in pressure requirements for sufficient airflow will often manifest itself as chest pain or discomfort, and the decrease in the number of alveoli clusters being supplied will generate the feeling of being short of breath.

In the worst case, the secretions and inflammation can actually end up damaging the alveoli reducing the ability of the lungs to exchange oxygen and other gases. Unfortunately, damaged alveoli cannot be repaired or replaced, so the damage and loss of lung function is permanent.

Like the common cold, bronchitis treatments often are aimed at alleviating the symptoms, not curing the disease. Depending on the cause, medications might include antibiotics or antiviral medications to treat any bacterial infection and expectorants which aid in making the cough more productive. In severe or chronic cases, an inhaler containing a bronchodilator might be used to increase the diameter of the bronchial passages. The inhaler or other medications might also include corticosteriods to address the inflammation of the brochial walls. If the lung efficiency is particularly bad, supplemental oxygen can be used to enrich the inhaled air providing a higher oxygen concentration to those parts of the lungs that are still working.

5.4   Asthma

On its surface, asthma looks very much like bronchitis. Both effect the bronchial tree and include swelling of the airway walls thereby narrowing the passageways as illustrated in Figure 5.4. Both present

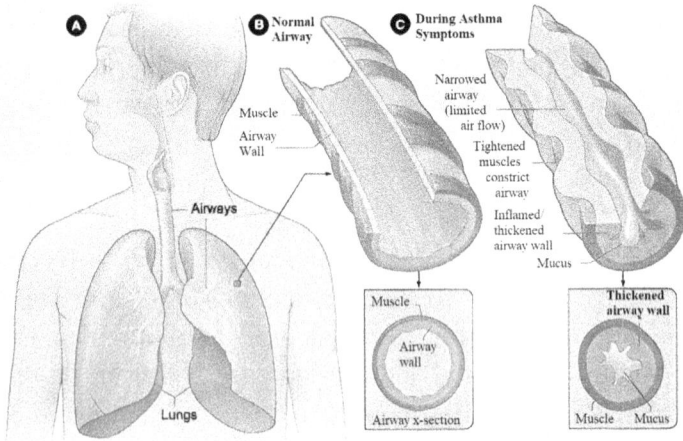

FIGURE 5.4

An asthmatic attack is caused by a sudden inflammation of the bronchial airways and can severely restrict and even stop the airflow to the alveoli. In the case of asthma, this is usually universal throughout the lungs. The onset is often rapid and caused by some triggering irritant, such as pollen, dust, food, smoke, or even stress. (National Institutes of Health, part of the United States Department of Health & Human Services.)

with coughing and wheezing that leads to shortness of breath and chest tightness. Asthma is also one of the three members of *COPD*.

However, there are key differences between asthma and bronchitis that make asthma an issue unto itself. For one thing, asthma is always chronic, whereas bronchitis can be either acute or chronic. That means that acute bronchitis can "go away" and be done never to reappear again. Asthma will not do that, it will always be present; however, the *symptoms* of asthma may disappear for days, weeks, months, or even years. On the other hand, chronic bronchitis, once you have it, will never go away, and neither will its symptoms. While both present with a cough, the cough of an asthmatic patient is typically dry. Asthma might present with inflamed airways, but there is little or no mucus production.

The asthmatic airway is always at least slightly constricted, which might pose an issue during strenuous exercise but otherwise be tolerable; however, it is not until an asthmatic attack that the constriction creates severe and possibly life-threatening issues. Furthermore, unlike bronchitis, the conditions of asthma are completely reversible, perhaps in just a few minutes for mild cases.

Asthma is often a genetic condition, something that is with the person from birth; however, early exposure to certain allergens as a young child or contracting certain lung diseases later in life may initiate the development of the asthmatic condition. Hazardous chemicals and pollutants in the workplace might also result in developing asthma.

Like bronchitis, the asthmatic attack is caused by a sudden inflammation of the bronchial airways and can severely restrict and even stop the airflow to the alveoli. In the case of asthma, this is usually universal throughout the lungs. The onset is often rapid and caused by some triggering irritant, such as pollen, dust, food, smoke, or even stress. There are many different triggering mechanisms, but they all lead to the same result. Attacks can last from minutes to days if not treated, and severity can also range from mild coughing and wheezing to fatal respiratory occlusion.

Rescue inhalers typically contain albuterol or levalbuterol which are bronchodilators. They alleviate the symptoms of asthma by causing the airways to open back up and allow airflow once again. The medication causes the muscles in the airway to relax. The elastic response then causes the airways to dilate making airflow easier. The effects of albuterol and levalbuterol are short-term, lasting only a few minutes to a few hours depending on dosage. Longer term preventative management typically uses corticosteroids.

5.5   Emphysema

In Section 4.6 we introduced the elastic modulus of the alveoli, but the nature and source of this elasticity were never discussed. The material that gives a great many tissues throughout the body their elastic properties is a substance known as *elastin* and comes in the form of fibers that connect across and cross-brace the tissues, providing them with both flexibility and structural integrity. These elastin fibers can be found in skin, ligaments, cartilage, blood vessels, and the lungs.

Emphysema is one of the three members of *COPD*, and it is caused when foreign particles make their way into the inner parts of the lungs, particularly the terminal bronchioles and alveoli. The body's response to these foreign particles, similar to its response anywhere else to a foreign entity like a splinter, is to lubricate the site of the irritation. Unfortunately, the lubricant that is generated inside the lungs frequently does little to eradicate the irritant, but it does permanently damage the integrity of the elastin fibers that give the terminal bronchioles and alveoli their elastic strength. The loss of the elastic integrity of the bronchioles turns them into limp hoses like so much overcooked bucatini. The alveoli also lose their integrity and end up collapsing on themselves as shown in Figure 5.5. Once this happens and the inner membranes of the alveoli come into contact with themselves, as discussed in Section 3.5.3, the pressure requirements to reopen those sacs becomes unrealizable, the sacs become useless, and the efficiency of the gas exchange process of the lungs is forever diminished.

Even if the alveoli remain open, the damage done to the terminal bronchioles that feed them can cause air to become trapped inside the connected alveoli. During the inhalation process, the tubes and alveoli are forced open as the chest wall expands and the pleural space between the wall and lungs drops below atmospheric pressure. However, when

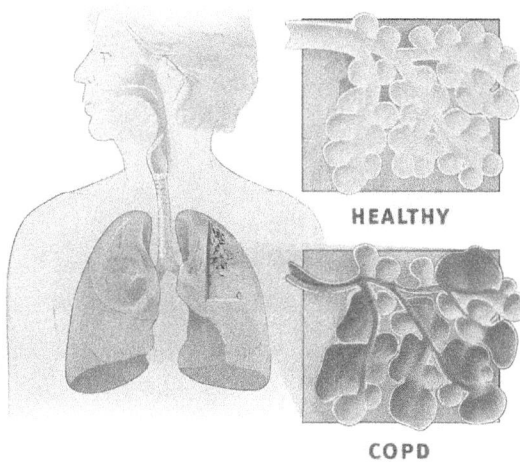

HEALTHY

COPD

FIGURE 5.5

Emphysema is one of the three members of *COPD*, and it is caused when foreign particles make their way into the inner parts of the lungs, particularly the terminal bronchioles and alveoli. Permanent damage is done to the elastic strength of the walls of the terminal bronchioles and alveoli. Consequently, the bronchioles have difficulty staying open, and the alveoli themselves also lose integrity. (National Institutes of Health, part of the United States Department of Health & Human Services.)

the chest wall moves inward, the pleural pressure becomes greater than the atmosphere as the chest attempts to drive the air out of the lungs. This increased pressure not only squeezes the air, but it also squeezes the entire lung. Under normal conditions, the elastic modulus of the alveoli and bronchioles would be sufficient to withstand that pressure, yielding only slightly; however, if the elastic integrity of the walls of the terminal bronchioles has become too severely compromised, the bronchioles can completely collapse, trapping the air inside their connective alveoli and preventing exhale.

To counteract this loss of integrity, patients experiencing advanced emphysema often resort to *pursed lips breathing,* whereby they purse their lips during exhale in order to restrict the airflow coming out of their lungs. Similar to the pressure drop through the larynx covered in Section 4.2, pursing the lips keeps the intrapulmonary pressures above atmospheric thereby keeping the pressure inside the alveoli and bronchioles higher than normal which assists in also keeping the tubes open and allowing the lungs to empty.

5.5.1   It Is Not About Bernoulli

We found in Section 4.4 that the mainstream velocity in the terminal bronchioles was approximately $v = 22.1 \ cm/s$. The pressure drop created by this velocity is given by Bernoulli's Equation 2.76 as

$$P = \frac{1}{2} \rho \, v^2 = 24.42 \ mPa \qquad (5.2)$$

As mentioned in 4.6, healthy elastin in the lungs have an elastic modulus of about 410 kPa. A 400 μm diameter terminal bronchiole has a circumference of $s = 1.257 \ mm$ and their walls are roughly $h = 1 \ mm$ thick. The elastic change in circumference due to the Venturi Effect can be found using Equation 2.31.

$$\Delta s = \frac{(24.42 \ mPa)(1.257 \ mm)}{410 \ kPa} = 74.85 \ pm \qquad (5.3)$$

which is negligible compared to the initial circumference. The change in thickness of the walls would be

$$\Delta h = \frac{(24.42 \ mPa)(1.0 \ mm)}{410 \ kPa} = 59.56 \ pm \qquad (5.4)$$

which is also negligible compared to the initial thickness. Clearly the Venturi Effect has little to do with the difficulties experienced by emphysema patients.

5.5.2   Pleural Pressure Effects

Normal relaxed breathing moves 500 mL of air 20 times per minute with a driving pressure of 5 $cm \ H_2O$ or about 490 Pa. A small portion of this pressure is due to the movement of the chest wall itself, but we will neglect that for this analysis, since such small effects are secondary to the desired physical understanding of *why* emphysema patients experience such difficulties and present such consistent and unfortunate breathing patterns.

Using the previous dimensions of the terminal bronchioles, an external pressure of 490 *Pa* would create a change of

$$\Delta s = \frac{(490\ Pa)\,(1.257\ mm)}{410\ kPa} = 1.502\ \mu m \tag{5.5}$$

in circumference and

$$\Delta h = \frac{(24.42\ mPa)\,(1.0\ mm)}{410\ kPa} = 11.95\ \mu m \tag{5.6}$$

in wall thickness. Again, these numbers are small with respect to the initial values, but this is healthy elastin. Under healthy circumstances, these changes are supposed to be negligible. Emphysema damages that elastic integrity and, eventually, might even destroy it entirely. If we knock off an order of magnitude, both of these numbers will also increase an order of magnitude. The walls of the bronchioles might be of most significance in this case, since healthy elastin already contributes an increase of an approximately 12 μm reduction in the original 400 μm inside diameter. Increasing that 12 μm an order of magnitude would occlude roughly 25% of the passageway. Clearly two orders of magnitude would occlude the passage altogether.

5.5.3   Stiffness and the Modulus

Of even higher significance is the connection between the elastic modulus and the *stiffness* of the bronchiole tubes, or alveoli for that matter. Granted, the surfactant lining the inside of the avleoli creates an amount of surface tension that is supposed to keep the alveoli open, but the loss of structural integrity of the wall on which that surfactant exists can make such surface tensions unimportant. Similarly, the terminal bronchioles can undergo a change in shape like running over a garden hose with a large truck. The hose circumference and wall thickness might not appreciably change in that situation, but the *shape* of the hose is now quite different and, if enough external pressure is applied, might collapse altogether and prevent any kind of flow through the pipe.

The elastic modulus is related to the stiffness of a pipe or tube in a manner very reminiscent of Hooke's Law in Equation 2.14 which was reviewed in Section 2.1.7. Consistent with the concept of Hooke's Law, the stiffness is given as k, where

$$k = E\left(\frac{\mathcal{A}}{L}\right) \tag{5.7}$$

where E is the elastic modulus, $\mathcal{A}$ is the cross-sectional area of the pipe, and L is the characteristic length of the pipe. Notice that the stiffness k has the same units as the spring constant, N/m. For a healthy bronchiole 1 *mm* in length, this equates to a stiffness of

$$k = (410\,kPa)\left(\frac{125.66 \times 10^{-9}\,m^2}{1\,mm}\right) = 51.5\,N/m \tag{5.8}$$

so it would take a force of

$$F = \frac{200\,\mu m}{51.5\,N/m} = 0.0103\,N \tag{5.9}$$

to "kink the pipe" and cut off the flow. A 1 *mm* long bronchiole has surface area of

$$\mathcal{A} = 1.2566 \times 10^{-6}\,m^2 \tag{5.10}$$

which would then require a pressure of

$$P = \frac{0.0103\,N}{1.2566 \times 10^{-6}\,m^2} = 8200\,Pa \tag{5.11}$$

to collapse the bronchiole tube.

One of the beauties of Physics is the ability to reverse the process and ask the question: how far can the elastic integrity deteriorate before the bronchioles collapse?

To answer this question, we start with normal breathing pressures of 5 *cm* H_2O of pressure and work backward. 490 *Pa* of pressure will apply a force of

$$F = (490\,Pa)\left(1.2566 \times 10^{-6}\,m^2\right) = 615.75\,\mu N \tag{5.12}$$

which implies a stiffness of

$$k = \frac{615.75\,\mu N}{200\,\mu m} = 3.0788\,N/m \tag{5.13}$$

and an elastic modulus of

$$E = \left(3.0788 \, \frac{N}{m}\right) \left(\frac{1 \, mm}{125.66 \times 10^{-9} \, m^2}\right) = 24.5 \, kPa \qquad (5.14)$$

which is just 6% of the original, healthy elastic modulus of $410 \, kPa$.

5.6 Pneumothorax/Hemothorax

As discussed in Section 3.5.1, the chest wall is not physically attached to the lungs. Between the wall and the lungs in a small region called the *pleural space* that provides a buffer between the chest wall and the lungs. The inner surfaces are lubricated to avoid damage or irritation of the either surface. During a normal respiratory cycle, this space undergoes both positive and negative relative pressures as the chest wall moves in and out, respectively. In this way, the wall generates a sympathetic response by the lungs causing them to expand or contract.

However, if something compromises the integrity of this airtight volume of space between the lungs and the wall, the chest wall will no longer be able to generate the positive and negative pressures required to garner a response from the lungs. Indeed, if the cavity fills with something other than a small amount of air and lubricant, the fixed volume that is supposed to be occupied by expanding lungs will be occupied by something else thereby reducing the available space for the lungs and thus the total lung capacity as shown in Figure 5.6.

Blunt force trauma to the rib cage can sometimes break two or more ribs in two places creating a "floating" section in the chest wall. The condition is known as a *flail chest*. Because the pressure inside the pleural space is negative to atmosphere on inhale, and positive to atmosphere on exhale, this floating section of ribs will actually move counter to the rest of the chest wall, being sucked in during inhale and bulging out on exhale. In addition to this moving segment being incredibly painful, the loss of wall integrity also impairs the breathing mechanics, and patients can experience considerable difficulty breathing.

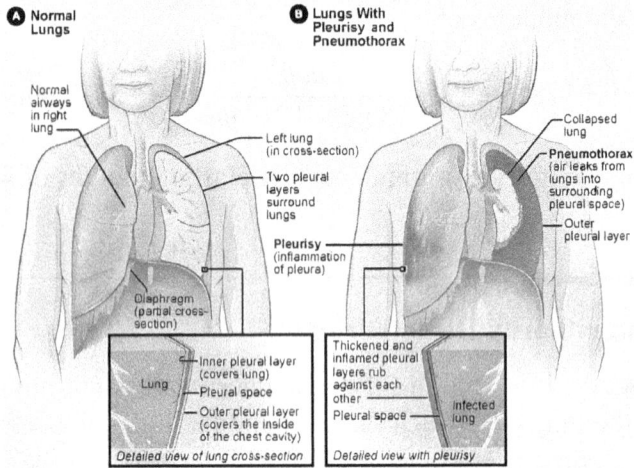

FIGURE 5.6

If the lungs become damaged either by disease or trauma, such as several alveoli rupture, the pleural cavity can fill with air. This is called a *pneumothorax*. Air can leak into the cavity with each breath. Little by little the pleural space fills with air slowly making it harder and harder to breath. (National Heart Lung & Blood Institute, part of the United States Department of Health & Human Services.)

If the lungs become damaged either by disease or trauma, such as several alveoli rupture, the pleural cavity can fill with air. This is called a *pneumothorax* which literally means air (pneumo) in the chest (thorax). Depending on the type of damage, this situation can be either acute or take place little by little as the patient breathes. Acute cases typically involve some type of trauma such as a traffic or industrial accident. Certain chemicals can cause immediate damage to the lung tissues causing air to go directly into the thoracic cavity. More long-term cases could involve a number of ruptured alveoli. The terminal bronchioles would then open directly into the pleural space, and a small amount of air would then leak into the cavity with each breath. Little by little the pleural space would fill with air slowly making it harder and harder to breath.

If the pleural space is filled more on one side than the other, as is often the case, the lungs would literally be pushed to one side over time. Since the lungs are hanging on the end of the trachea, the airway would move along with the lungs. This condition is known as a *deviated tra-chea* and signifies that the patient will require immediate intervention. At this point, it is unlikely that the patient has sufficient lung capacity to survive for very long. Anywhere from 30% to 60% of the thoracic cavity has been compromised, and the lungs are being squeezed into a corner. This advanced and life-threatening condition is known as a *tension pneumothorax.*

A similar situation can occur if blood vessels in the thoracic region are damaged or compromised in such a way that causes bleeding into the pleural space. This condition is called a *hemothorax,* meaning blood (hemo) in the chest and is typically caused by some sort of trauma to the chest area, but it can also be caused by chemicals or diseases that damage or deteriorate the integrity of the walls of the blood vessel. The rate at which the cavity fills is directly tied to the type and size of the artery or vein and the size of the "leak." Large arteries such as the aorta and pulmonary arteries, or large veins such as the vena cava and pulmonary veins, can dump a lot of fluid into the pleural space in a short time; whereas, small capillaries will just drip little by little and take hours or days to produce noticeable symptoms. Just as a pneumothorax can advance into a *tension pneumothorax,* significant amounts of blood in the chest cavity can also produce a *tension hemothorax* in the same way. In this case, the patient is not only suffering from a significant lack of lung capacity but probably also from serious blood loss, or *hypovolemia.*

Treatment or correction of a pneumothorax or hemothorax start with removing the occupying fluid, either air or blood, and allowing the lungs to expand. This can be done by injecting either a needle (needle aspiration) or a larger chest tube allowing the air or blood to escape. The lungs would then be free to expand to their normal volume once again.

The second issue is repairing the leak that caused the problem in the first place. In the case of the pneumothorax, that means plugging the air leak. There are a number of non-surgical treatment options available that can create a "plug" over the leak and prevent air from getting into the pleural cavity again. If the situation is severe, it may require surgical intervention and possibly removal of the damaged section.

In the case of the hemothorax, repairing the bleeding can be a bit more straightforward. The body is normally pretty good at addressing "leaks" in its circulatory system, and sometime just giving it time is all that is needed. In this case, the drain tube (thoracotomy) is left in place for a few days to allow the body time to "clot." Medications can be given that will help the body develop the clots. If the bleeding is persistent or severe, surgical correction may be required to repair, tie off, or cauterize the damage blood vessels.

5.7   Respiratory Syncytial Virus

Respiratory Syncytial Virus, or *RSV*, is a common respiratory virus that typically causes mild, cold-like symptoms. While most people recover from RSV within a week or two, it can be serious, especially for infants and older adults, leading to severe lung infections and pneumonia. *RSV* is a highly contagious virus that spreads easily through respiratory droplets. The symptoms of *RSV* vary depending on the age and health of the individual, but in older children and adults the symptoms are often similar to those of the common cold.

RSV is a virus that belongs to the family *Pneumoviridae*. It is one of the most common causes of respiratory illness in infants and young children worldwide. Almost all children will have had an *RSV* infection by their second birthday. While often mild, *RSV* can lead to significant health issues, particularly in vulnerable populations. The virus primarily infects the cells lining the respiratory tract, from the nose and throat down to the lungs. In severe cases, it can cause inflammation of the small airways in the lungs and infection of the air sacs leading to viral pneumonia.

5.8   Cystic Fibrosis

Cystic Fibrosis is a genetic disorder that primarily affects the lungs, but it can also affect the pancreas, liver, kidneys, and intestines. It is characterized by the production of thick, sticky mucus that can clog airways and obstruct ducts in various organs. This unusually abnormal mucus is the hallmark of Cystic Fibrosis and leads to the wide range of symptoms and complications associated with the disease.

The most significant impact of cystic fibrosis often manifests itself in the respiratory system. The thick mucus in the lungs traps bacteria and other pathogens leading to recurrent and chronic lung infections. These infections, combined with inflammation, cause progressive lung damage, including a widening and scarring of the airways and eventually respiratory failure, which is the leading cause of death in patients suffering from the disease.

The digestive system can also be severely affected. The thick mucus can block the ducts coming from the pancreas preventing digestive enzymes from reaching the small intestines. This malabsorption leads to nutrient deficiencies, low weight gain, and other digestive issues. Many individuals require enzyme replacement therapy to aid in the digestion of foods and in promoting nutrient absorption. The liver may also develop blockages in its bile ducts resulting in cirrhosis or other liver damage. The sweat glands can start to produce sweat with abnormally high salt content, which is a key diagnostic indicator.

While there is currently no cure for cystic fibrosis, significant advancements in treatment have dramatically improved the quality of life and life expectancy for individuals with the condition. Treatment strategies such as chest physical therapy, vibrating vests, and breathing exercises focus on managing symptoms, preventing complications, and improving overall health. Medications may include antibiotics to treat and prevent lung infections, anti-inflammatory drugs to reduce airway inflammation, bronchodilators to open airways, and mucolytics to thin out the mucus. Pancreatic enzyme supplements are also essential for digestive health.

In recent years, a revolutionary class of drugs called *CFTR* modulators has emerged that target the underlying genetic defect. Depending on the specific mutation, these modulators can significantly improve lung function, reduce exacerbation, and enhance overall health, representing a major breakthrough in care. Despite these advancements, managing cystic fibrosis requires a lifelong dedication and multidisciplinary approach. Patients typically receive care from a specialized

team of healthcare professionals, including pulmonologists, gastroen-
terologists, dietitians, social workers, and physical therapists, all working together to address the complex needs of the individual. Ongoing
research continues to explore new therapies, including gene editing
and other novel approaches, with the ultimate goal of finding a cure for
cystic fibrosis.

5.9   Tuberculosis

Tuberculosis, commonly called *TB*, is an infectious disease caused
by the bacteria called *mycobacterium tuberculosis*. It was the leading
cause of death worldwide for centuries earning grim monikers like
"consumption" due to its debilitating effects on the body. While often
associated with the lungs, *TB* can affect any part of the body including
the kidneys, spine, and brain leading to a wide range of symptoms and
complications.

The transmission of TB primarily occurs through tiny airborne
particles containing the bacteria released into the air when an infected
person coughs, sneezes, or speaks. These particles are then inhaled by
others leading to infection. It is important to note that not everyone
exposed to the bacteria will develop the disease; many will develop a
latent infection where the bacteria remains dormant in the body without
causing any symptoms. Latent *TB* infection is a critical aspect of the
disease's epidemiology. Individuals do not feel sick, do not have any
symptoms, and cannot spread the bacteria to others. However, they are
at risk of developing an active case at any time in the future, especially
if their immune system becomes weakened. This makes contact tracing
and preventive treatment crucial in controlling the spread of the disease.

When a latent *TB* infection progresses to active *TB* disease, the
symptoms typically include a persistent cough often with blood-tinged
sputum, chest pain, weakness or fatigue, weight loss, fever, and night
sweats. These symptoms can be subtle at first, leading to delays in both
diagnosis and treatment which can increase the risk of transmission to
others.

Treatment for active *TB* disease is a lengthy and complex process, typically involving a multi-drug regimen for at least six months. Drug-resistant *TB*, poses a significant challenge to global health as some strains are resistant to at least one of the two most powerful anti-*TB* drugs. The emergence of drug resistance is often linked to improper or incomplete treatment regimens, as well as inadequate infection control measures.

5.10 Lung Cancer

Lung cancer stands as one of the most prevalent and aggressive forms of cancer globally. As the name suggests, lung cancer originates in the cells of the lungs. It is characterized by the uncontrolled growth of abnormal cells, cancer cells, which can form tumors and, if left untreated, spread to other parts of the body through a process known as *metastasis*. This disease poses a significant public health challenge, being a leading cause of cancer-related deaths worldwide.

The two primary types of lung cancer are small cell and non-small cell lung cancer. Of the two, non-small cell lung cancer is the more common type accounting for approximately 85% of all lung cancer cases. It can manifest itself as adenocarcinoma, squamous cell carcinoma, and large cell carcinoma, each presenting with its own distinct cellular characteristics and growth patterns. Small cell lung cancer, while less common, is typically more aggressive and tends to spread more rapidly.

Lung cancer cases are often linked to the inhalation of some type of particulate carcinogen that damages the *DNA* in the lung cells leading to mutations that can ultimately trigger cancerous growth. These carcinogens might include radon gas, asbestos, certain industrial chemicals, air pollution, smoke, or a family history of lung cancer.

In its early stages, lung cancer often presents with no noticeable symptoms, making early detection challenging. As the disease progresses, individuals may experience persistent cough, shortness of breath, chest pain, wheezing, hoarseness, unexplained weight loss, fatigue, and recurrent infections like bronchitis or pneumonia. These symptoms are often mistaken for other less serious conditions, further delaying both diagnosis and treatment.

Diagnosis of lung cancer typically involves a combination of imaging tests, such as chest X-rays, CT scans, and PET scans, to identify suspicious areas in the lungs. If an abnormality is found, a biopsy is performed to collect tissue samples for pathological examination, confirming the presence and type of cancer. Once lung cancer is diagnosed, it is staged to determine the extent of the disease. Staging helps classify the cancer based on the size of the tumor, whether it has spread to nearby lymph nodes, and if it has metastasized to distant organs. This information is crucial for guiding treatment strategies and predicting prognosis.

Treatment for lung cancer is highly individualized and depends on the type, stage, and overall health of the patient. Common treatment modalities include surgery to remove the tumor, chemotherapy to kill cancer cells, radiation therapy to target and destroy tumors, and targeted therapy drugs that specifically attack cancer cells with certain genetic mutations. Immunotherapy, which boosts the body's own immune system to fight cancer, has also emerged as a significant advancement in recent years.

Prevention remains the most effective strategy against lung cancer. Avoiding exposure to harmful particles, testing homes for radon, and taking appropriate precautions in occupational settings where carcinogens are present are vital. The prognosis for lung cancer varies widely depending on the stage at diagnosis and the effectiveness of treatment, but early detection significantly increases the chances of successful treatment and long-term survival.

Living with lung cancer can be a challenging journey, impacting not only physical health but also emotional and mental well-being. Ultimately, the importance of early detection cannot be overstated.

Bibliography

[1] Samer Bou Jawde, Ayuko Takahashi, Jason H. T. Bates, and Béla Suki. An analytical model for estimating alveolar wall elastic moduli from lung tissue uniaxial stress-strain curves. *Frontiers in Physiology*.

[2] Daroszewski, M., Szpinda, M., Flisiński, P., Szpinda, A., Woźniak, A., Kosiński, A., Grzybiak M. & Mila-Kierzenkowska C. Daroszewski, M. Tracheo-bronchial angles in the human fetus – an anatomical, digital, and statistical study.

[3] Balasubramanian S., Kalaskar A. and Kalaskar, R. Evaluation of the average nasal and nasopharyngeal volume in 10–13-year-old children: A preliminary cbct study.

[4] Nyengaard J. R., Jung A., Knudsen L., Voigt M., Wahlers T., Richter J. & Gundersen H. J., Ochs M. The number of alveoli in the human lung.

[5] I. R. Titze. Physiologic and acoustic differences between male and female voices.

[6] Welch B., Gold K., Gartner-Schmidt J., Petrov A., Law A., & Helou L. B. Variability of maximum glottal angle on clinical sniff task differs in patients with functional and organic laryngeal pathologies compared to healthy controls. *Journal of Voice*, 2024.

[7] Zhang Z. Contribution of laryngeal size to differences between male and female voice production.

Index

For Product Safety Concerns and Information please contact our EU
representative GPSR@taylorandfrancis.com
Taylor & Francis Verlag GmbH, Kaufingerstraße 24, 80331 München, Germany